T0213618

Finite Element Analysis of Weld Thermal Cycles Using ANSYS

Finite Element Analysis of Weld Thermal Cycles Using ANSYS

G. Ravichandran

General Manager (Retd)
Welding Research Institute
BHEL, Tiruchirappalli
Professor Adjunt
NMAM Institute of Technology
Nitte, Karnataka

CRC Press
Taylor & Francis Group
Boca Raton London New York

CRC Press is an imprint of the
Taylor & Francis Group, an informa business

First edition published 2021
by CRC Press
6000 Broken Sound Parkway NW, Suite 300, Boca Raton, FL 33487-2742

and by CRC Press
2 Park Square, Milton Park, Abingdon, Oxon, OX14 4RN

ISBN: 978-0-367-51019-0 (hbk)
ISBN: 978-0-367-53576-6 (pbk)
ISBN: 978-1-003-05212-8 (ebk)

Typeset in Times
by Lumina Datamatics Limited

Contents

Preface...vii
Author ix

Chapter 1   Introduction.. 1

Chapter 2   Arc Welding Processes...5

2.1   Electric Arc 5
2.2   Classification of Welding Processes..................................7
2.3   Welding Power Source 8
2.4   Welding Techniques 10
2.5   Shielded Metal Arc Welding Process 11
2.6   Submerged Arc Welding Process..................................... 12
2.7   Gas Tungsten Arc Welding Process............................... 13
2.8   Gas Metal Arc Welding Process..................................... 15

Chapter 3   Thermal Cycles and Heat Flow in Welding 17

3.1   Heating and Cooling Cycles... 17
3.2   Heat Flow in Base Metal.. 18
3.3   Graphical Plotting of Results 19
3.4   Factors Influencing Thermal Cycles 22
3.5   Simplification of Three-Dimensional Model 23

Chapter 4   Finite Element Analysis 25

4.1   Shape Function..25
        4.1.1   Shape Function for a One-Dimensional Simplex
                   Element..25
        4.1.2   Shape Function for a Four Noded Quadrilateral
                   Element..27
4.2   Formulation of Equation 29
4.3   Finite Element Analysis Using ANSYS....................32

Chapter 5   Arc Heat Model..35

5.1   Cross Sectional Analysis...39
5.2   In-Plane Analysis 49
References 107

Chapter 6   Sample Problems...109

      6.1   Cross Sectional Analysis of a Submerged Arc
           Welded Plate 109
      6.2   In-Plane Analysis of a Gas Metal Arc Welded Plate.............126
      6.3   In-Plane Analysis of a Gas Metal Arc Welded Dissimilar
           Weldment 144
      6.4   In-Plane Analysis of a Gas Metal Arc Welded Plate.............167
      6.5   Three-Dimensional Analysis of a Gas Tungsten Arc
           Welded Tube...186

Chapter 7   Conclusion 207

Appendix: Exercise Problems..209

Bibliography 221

Index..223

Preface

Finite element method is a powerful tool which can be employed for analyzing various aftereffects associated with welding. The knowledge can help a practicing engineer to arrive at the optimum procedure which will minimize the extent of rework in critical components. But getting a meaningful result using FEM is no easy task, and it requires a great amount of effort to create input data which truly represent the important phenomena associated with welding. An equal amount of effort may be required to sift through the vast data output and interpret the results. Thus the engineer must have a good knowledge of the phenomena associated with welding as well as the mathematical skill to express them as input data.

This book has been written to educate young researchers on the analysis of thermal cycles during the welding process. The related inputs on the welding processes, finite element method, the calculation of arc heat input for different cases and example problems with input data and typical output have been presented sequentially in the chapters of the book. The example problems have been solved using ANSYS software, which is one of the popular general purpose softwares for the analysis of engineering components.

The introduction chapter presents the various metallurgical and mechanical aftereffects of welding succinctly. The chapter on arc welding processes describes the arc phenomenon, the different arc welding processes and welding techniques, typical welding parameters employed in each welding process, heat transfer efficiency for various processes, etc.

The thermal cycles which are associated with welding are discussed in the third chapter. The conduction, convection and radiation heat transfer modes, heat flow in the base metal, the heating and cooling cycles, graphical representation of results, such as temperature versus time plots, isothermal plots and temperature distribution with distance are described. Procedures to simplify the complex three-dimensional analysis into an equivalent two-dimensional analysis are also explained.

The fourth chapter gives basic information on the finite element method, the calculation of shape functions for one-dimensional and two-dimensional cases, the formulation of the problem and the solution. The procedure for running of the analysis using ANSYS software is also explained.

The mathematical representation of the arc heat and the procedure to calculate the input data for the FEM program for different cases have been explained in detail in the fifth chapter. The examples are selected in such a way that the reader will get a good understanding and apply the knowledge for running his own problem.

The sixth chapter presents example problems and their solutions describing the procedure to prepare the input data for the ANSYS program from the calculation of arc heat described in the previous chapter. The cross sectional analysis, in-plane analysis and a full three-dimensional analysis are covered in the examples.

The knowledge for writing the book was acquired during my long years of work experience in Welding Research Institute, Bharat Heavy Electricals Limited (BHEL), Tiruchirappalli. The conducive environment for learning, support from the superiors and the freedom at work were the key factors which helped me in my learning, and I extend my heartfelt gratitude to the management of WRI, BHEL, Tiruchirappalli for the same. I thank the management of the NMAM Institute of Technology, Nitte for the encouragement during the preparation of the book. I also extend my thanks to M/s ANSYS Inc. for permitting me to reproduce ANSYS input files and output plots.

Author

G. Ravichandran had 37 years of experience with the Welding Research Institute, which is affiliated to BHEL, Tiruchirappalli, India, and retired from WRI in 2017. His field of specialization is welding simulation using finite element method, prediction and control of distortion and residual stresses, design of static and fatigue loaded welded structures, experimental stress analysis, etc. He obtained his PhD from Indian Institute of Technology, Madras in 1996 for his work on the analysis of axis shift distortion and residual stresses during circumferential welding of thin walled pipes. He received the W. H. Hobart memorial award from the American Welding Society in 1997 for his research paper "Prediction of Axis Shift Distortion during Circumferential Welding of Thin Pipes Using the Finite Element Method." He was a guest faculty for many educational institutes and handled classes for both graduate and postgraduate programs on welding engineering. After his retirement, he joined NMAM Institute of Technology, Nitte, Karnataka, as an adjunct professor in the mechanical engineering department.

1 Introduction

Welding is widely employed in the fabrication industry for the manufacture of many types of components involving different shapes, sizes, and materials. The main advantage of the welding process is that it produces a metallurgical bond between the members to be joined which has a strength level comparable to or even higher than that of the base material of the component. Before the advent of the welding process, riveting was the main metal joining process for the fabrication of components such as structures, ships and even some pressure vessels. Riveting has several advantages over welding like lower equipment cost, simple operation, lower operator skill level, standardized procedure for different materials without any additional operations like preheating and post-weld heat treatment, less consumable requirement, simple quality control procedures, etc. But the advantage of higher joint strength in the case of welding gives it an edge over riveting in many types of components. After the development of the arc welding process, which progressively produces coalescence of metals along the joint line by a continuous supply of energy, welding has found widespread industrial usage. The welding process enables the designer to optimally design the component for maximum performance with minimum size, shape and weight of the component. Today, industrial components are designed with advanced materials and thicker sections for operating at increasingly higher load levels. The challenges in the joining of these critical components are effectively met by welding.

Much research work has been undertaken to address the various issues which are faced during welding, and the welding process owes its premier position to these developments. Many welding processes, procedures, consumables and equipment have been developed which are suitable for various types of components for different end applications. The productivity of the welding process is continually being improved, and automation has been successfully implemented for large-scale manufacture of many welded components. Several studies have been undertaken to understand the metallurgical issues during the welding process and the ways to overcome them. Suitable welding procedures which minimize the metallurgical damage in different materials are under constant development. The deleterious effect of weld discontinuities on the load carrying capability has been studied in detail, and steps to overcome the problems are widely known to the industry. Advanced testing methods have been developed for better reliability of the welded components. Fabrication codes have been formulated so that the welding procedure is standardized and safe performance of welded components during service is ensured. There are many trained welders and welding operators who meet the skill requirement of various fabrication codes. Thus the process of welding has come to occupy an important position in the fabrication industry.

The metallurgical bond in the case of welding is achieved by the conversion of different forms of energy into the energy for coalescence. This energy is focused in the localized region at the interface between the two members. In the case of arc welding processes which account for a major share of welding, the electric arc provides the required energy for fusion of the members. The arc almost instantaneously raises the temperature of the base metal beyond the melting point. In many cases, extra metal is deposited to fill the groove between the members, and this additional metal also receives the heat from the electric arc. After solidification, the required metallurgical bond between the abutting members is created.

The success of the welding process depends on the localized liberation of heat along the joint line, but this localized heat has its own aftereffects and unless these detrimental aftereffects are suitably tackled, the resulting quality of the weld will not be satisfactory. The zone in the parent metal which is closest to the weld is heated to very high temperature, which causes metallurgical phase changes in the material. This zone which is metallurgically affected by the high temperature is called the heat affected zone. This zone is the weakest link in the entire weldment, and unless steps are taken to contain the weakness within limits, the resulting weld will not meet the quality requirements.

The heat due to welding produces different effects in different materials under different conditions. In materials like carbon steel and low alloy steel, it produces hardening due to rapid cooling whereas in materials like aluminium alloys, it has a softening effect due to recrystallization which takes away the beneficial effects of cold work. In steels, faster cooling rates during welding affect ductility whereas slow cooling rates affect toughness. The welding heat causes grain refinement in the previously deposited weld metal in multipass welding whereas it has a grain coarsening effect in the heat affected zone which affects the strength and toughness. The welding heat leads to depletion of elements like chromium, which give the required properties, whereas it may cause segregation of impure elements, which weakens the weld joint. In stainless steel, the welding heat may lead to the precipitation of unfavourable carbides, and in some other cases, the heat leads to dissolution of the favourable carbides in the matrix. The inadequate liberation of the welding heat results in improper fusion of the weld metal with the base metal but in some steels, the excess liberation of heat causes melting of more parent metal which alters the chemical composition of the weld metal and makes the weld metal inferior.

The non-uniform heating and cooling cycle associated with welding also causes problems of residual stresses and distortion which affects the integrity of the welded structure. The differential thermal expansion during the heating cycle causes localized plastic deformation in the weld and adjacent regions, and as a result, shrinkage forces are developed in this region during the cooling phase. These shrinkage forces produce residual stresses or distortion depending on the restraint offered by the component. In steels containing non-metallic inclusions in the form of thin lamella at the mid-section, the shrinkage forces in the thickness direction may lead to tearing of the member, and this phenomenon is known as lamellar tearing.

The root cause of all these detrimental aftereffects can be traced to the application of localized heat during welding, and hence the total elimination of these aftereffects is not possible. But the damage may be contained within limits by controlling the heat input to the exact minimum requirement. Any indiscreet application of heat during welding will aggravate the problem, and post correction becomes more difficult. Thus the successful application of welding requires a careful balancing of the heat input for achieving proper fusion without severe degradation of the properties of the weld.

The knowledge of the temperature history during the welding process is the essential first step for understanding the metallurgical and mechanical aftereffects due to welding. There are analytical methods for the study of temperature history in the weldment. These methods are easy to use and have been successfully employed in some cases. But these methods employ many simplifying assumptions which affect the accuracy of the results.

With the improvement in computing power, numerical techniques like the finite element method have become popular as various phenomena may be incorporated easily in these models and reasonably accurate thermal results can be obtained. There are many phenomena associated with welding and accounting all these phenomena will make the analysis very complex and time consuming. The phenomena which have major effects must be considered in the model, leaving out the other less important ones so that the accuracy of the results will be reasonable while the cost of computation is kept within limits. Incorporating the phenomena in the model must be done carefully, and improper representation of these phenomena will have adverse effects on the accuracy of the results.

There are commercially available finite element method based softwares which can help a researcher to perform the analysis without the need to go through the basic steps in formulation and solution of the problem. These softwares enable the user to visualize the results for better understanding. Even though there are many advantages in using these softwares, the user has to exercise caution during the data input since there may be some underlying limitations for a particular case, and if the researcher tries to run the analysis without really understanding the effect of these limitations, it will only lead to inaccurate and unreliable results.

This book has been written to educate the fresh researchers about the analysis of thermal cycles in welding using ANSYS software. The reason for considering ANSYS software in this book is the familiarity of the author with the software. Other softwares which have similar features as ANSYS may also be employed for the analysis provided the inputs as discussed in this book are suitably entered.

2 Arc Welding Processes

The welding processes can be broadly classified as fusion welding and pressure welding processes. In fusion welding, the abutting edges of the members are melted and the thoroughly mixed common molten pool after solidification forms the metallurgical bond between the members. For joining thick sections, additional metal has to be supplied to the molten metal pool for filling up the gap between the members, and this additional metal mixes thoroughly with the molten base metal before solidification.

The other type of welding process uses the application of pressure across the interface to dislodge the surface impurities and bring the fresh metal surfaces within interatomic distance. The base metals in this case may or may not be heated during the pressure application, but the temperature, in any case, does not exceed the melting point of the base metal. Hence, these processes are also referred to as solid phase welding processes. The solid phase welding processes have many advantages over fusion welding processes such as feasibility to weld dissimilar combination of materials, very narrow heat affected zones, etc. But these processes are limited to simple geometries and hence, have application for only a small range of products.

The energy for the formation of metallurgical bond between the base metals is obtained by the conversion of different forms of energy such as chemical energy, electrical energy, beam energy, etc. Arc welding processes are the most common type of fusion welding processes which are widely employed in industry. The processes have a common feature that the energy for the fusion of the abutting edges of the two components is provided by an electric arc. The arc is an electric discharge which generates intense heating that is used for fusion of the base metals. There are different types of arc welding processes which cater to different applications.

2.1 ELECTRIC ARC

The electric arc phenomenon was discovered by Sir Humphrey Davy. In an electric circuit with two carbon electrodes which are connected to the terminals of a power source, an electric current flows when the two electrodes are made to come in contact with each other. But subsequently when the two electrodes are slightly retracted so that there is a small gap between the two electrodes, an electric discharge takes place between the electrodes which ionizes the gaseous matter between the electrodes. This electric discharge is called *arc* and the electric circuit is completed by the movement of the ions and electrons in the arc column which move towards the negative and the positive terminals, respectively. Unlike the phenomenon of **spark** which involves high voltages and lower current levels, the arc phenomenon involves lower voltages and higher current levels. The high voltage in a spark enables it to jump across large gaps, but the voltage level in the arc phenomenon is lower and the resulting arc gap is also less. If the gap between the electrodes is increased beyond some limit, then the arc gets extinguished.

The carbon electrode arc is rarely used for welding as the heat energy can hardly be focused towards the base metal and the fraction of energy received by the base metal is insufficient for complete fusion of the metal especially when the thickness is high. On the other hand, if the arc is created between a metal electrode which is in the form of a rod and the base metal, then the arc is directed towards the base metal and hence, the heat released in the arc creates a large molten pool in the base metal. The electrode is also heated by the arc and in the case of consumable electrode, the arc heat melts the tip of the electrode and the molten metal from the electrode tip is transferred across the arc to the base metal. This provides the filler metal which is required to fill the groove between the base metals.

The arc in the case of metal electrode takes the shape of a bell and diverges from the tip of the electrode to the base metal. The diverging nature of the arc with decreasing levels of current density at various sections results in the creation of magnetic forces. The radial component of the magnetic force, which is called the pinch force, is responsible for the detachment of the molten metal droplet from the tip of the consumable electrode. The other axial component is directed towards the base metal and is responsible for the transfer of the molten metal across the arc to the molten pool below the arc thus creating a digging effect into the base metal. The bell-shaped arc covers a small area in the base metal and the energy generated in the arc is transferred to the base metal through this small area.

The electrical parameters like current and voltage have a major influence on the shape and nature of the arc. The voltage across the arc depends on the gap between the electrode and base metal which is known as arc length. As the arc length increases, greater area is covered by the arc and consequently the width of the resulting weld bead is higher. The current which flows through the arc decides the energy of the electrons in the arc column and hence, the extent of fusion in the base metal is decided by the current. The heat generated in the arc column is calculated as the product of current and voltage.

Even though the arc is struck directly over the base metal, not all the heat generated in the arc is transferred to the base metal. Some energy is lost to the environment by convection and radiation heat loss, heat loss to the cooling medium if any, heat carried away by metal particles thrown away in the form of spatter and vaporization of the metal. The actual heat realized in the base metal is less than the power generated by the arc, and the ratio of the heat energy received by the base metal to the energy released in the arc is termed as efficiency of the welding process. The process efficiency is dependent on the nature of the arc welding processes and the particular welding conditions. The exact value of the process efficiency always lies within a range and has to be selected for a particular condition carefully.

In the case of consumable electrode welding processes, the arc heat causes melting of the tip of the electrode and the molten metal droplet is transferred across the arc. This melting of the tip of the electrode results in increase in the arc length, and in order to prevent the arc from getting extinguished, the electrode must be moved towards the base metal. This progressive movement of the electrode towards the base metal is known as electrode feeding. In consumable electrode welding processes, the electrode feed rate decides the current and the metal deposition rate.

After striking the arc and after steady operating parameters are reached, the arc is steadily moved along the joint. The uniform movement of the arc along the joint ensures a uniform weld along the entire length. The various points lying on the joint line receive the heat from the arc only for a short duration, but the intense heat generated by the arc results in near instantaneous melting of the base metal. The duration for which any point receives the arc heat depends on the welding speed, and hence the heat energy liberated at any point is directly proportional to the power of the arc and inversely proportional to the speed of welding. The heat energy liberated per unit length of the weld is given in the unit of joules per mm (J/mm) or kilojoules per mm (kJ/mm). It is calculated as $\eta.V.I/s$ where η is the welding process efficiency, V is the voltage, I is the current, and s is the welding speed in mm/s.

2.2 CLASSIFICATION OF WELDING PROCESSES

The major obstacle to overcome in the use of electric arc for welding is the effect of atmospheric gases on the molten metal. When the high temperature metal in the weld zone comes in contact with the atmospheric gases, the gases may get dissolved in the molten metal or may chemically react with the high temperature metal. This results in the formation of solid or gaseous inclusions in the deposited metal which affects the integrity of the weld joint. Hence, for getting sound weld metal which is free from inclusions, the arc has to be protected against the action of the atmospheric gases. The protection is given either by the use of a chemical substance called flux or by the use of an inert gas. Thus the welding processes are classified as flux shielded welding processes or gas shielded welding processes.

The welding process involves the movement of the arc along the weld line, and this movement of the arc may be performed manually by the welder or automatically by the use of a device. Thus, the arc welding processes are classified as manual welding or automatic welding. Automatic welding with uniform deposition of weld metal is ideally suited for filling up deep grooves in long weldments. But in places like root pass welding where the gap between the two abutting edges may be non-uniform, the manual process has an edge over the automatic process. The welder in such cases of uneven gap can adjust the welding speed so that the resulting weld bead smoothly merges with the base metal. The skill of the welder plays a major role in the appearance of the weld bead.

The arc is struck between the metal electrode and the base metal, and the heat of the arc is realized both at the base metal end and the electrode end. In most cases, the electrode also melts under the arc heat and gets consumed. It is also possible to carry out welding by striking an arc between a tungsten electrode and the base metal. The electrode in this case does not melt due to the arc heat and the additional metal for filling up the groove has to be supplied using an externally fed filler metal in the form of rod or wire. Thus, there is one more classification of the welding processes, namely consumable electrode welding processes and non-consumable electrode welding processes.

2.3   WELDING POWER SOURCE

The electric power for welding comes from a power source which delivers the welding current at the required voltage level. The required voltage for sustaining the arc will be in the range of 10–40 V and the welding current is usually in the range of 50–600 A. The power sources are designed to deliver the required current levels either continuously or intermittently. In the case of automatic welding processes, a continuous rating of the power source is desired whereas for manual welding process with coated electrodes, an intermittent rating is sufficient as the welding process is discontinuous by nature.

The welding power sources are designed in such a way that either the current or voltage is kept nearly constant throughout the process while the other parameter is permitted to vary over a wide range. It was seen earlier that the voltage is influenced by the arc length whereas the current influences the melting rate of the electrode. Under stable operating conditions, the electrode feed rate is equal to the electrode melting rate and the arc length is maintained constant. But some small disturbance in the arc length is inevitable in both automatic and manual processes, and this disturbance will affect the welding parameters. Unless suitable corrective steps are taken, the resulting weld bead appearance will not be good.

In the case of an automatic welding process, the constant voltage characteristic power source as shown in Figure 2.1 is usually used. Under stable operating conditions, the power source delivers a current and voltage level. Any disturbance to the arc, which results in an increase of the arc length, leads to a higher voltage level than what was set earlier. Due to the power source characteristics of constant voltage, an increase in voltage level results in reduced current levels, which correspondingly reduces the electrode melting rate. This reduction in electrode melting rate when the feed rate is maintained at the original level restores the arc length to the originally set value. Similarly, any instantaneous reduction in arc length leads to higher current levels, and the consequent higher melting rate restores the arc length to the original value. The correction is usually very quick without affecting the overall melting rate or the extent of fusion in the base metal. This is known as the automatic voltage correction mechanism.

This type of power source cannot be employed in the case of a manual process in which greater variation in arc length is expected even for experienced and skilled

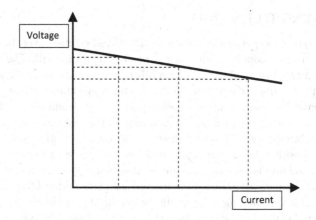

FIGURE 2.1   Constant voltage characteristics of welding power source.

welders. A quick correction in the arc length cannot be achieved as the current densities in manual cases are not high like in automatic welding processes. Hence, the power source for manual welding should be such that the frequent variation in arc length should not affect the melting rate of the electrode. For the manual welding process, a power source which delivers nearly constant current levels irrespective of the disturbances to the arc length is needed. The characteristics of this type of power source are shown in Figure 2.2. The weld bead appearance depends to a large extent on the skill of the welder.

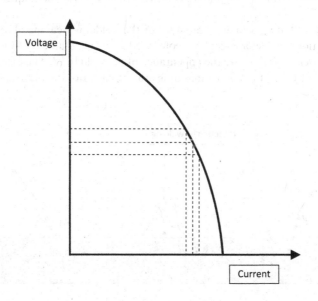

FIGURE 2.2   Constant current characteristics of welding power source.

2.4 WELDING TECHNIQUES

The welding process is performed mostly in downhand position over the base metal from the top. In this condition, the welder needs very little skill. The welder can deposit large amount of weld metal, which fills the groove due to the action of gravity. But not all welding is carried out in the downhand position. In many instances, the welder has to perform welding in other positions also, and the welder needs a special skill level to deposit the weld metal against the force of gravity. The other positions in which the welder has to perform are horizontal position, vertical position and overhead position. In the horizontal position, the base metal lies in the vertical plane and the welding is performed along the horizontal groove between the plates. In the vertical position, the plates are in the vertical plane and welding is performed along a vertical line from bottom to top. In the overhead position, the base metal lies in the horizontal plane, but the weld metal has to be deposited from the bottom side.

Welding is usually carried out with the electrode being held perpendicular to the base metal. But welding can also be carried out with the electrode being tilted along the direction of welding. This has led to the development of forehand technique and backhand technique apart from the normal technique. In the forehand technique, the electrode is tilted in the forward direction so that the arc is directed towards the base metal ahead of the molten pool. In this case, the base metal is heated by the arc and the metal is deposited later. This results in deeper extent of penetration of the parent metal. In the backhand technique, the electrode is tilted backwards so that the weld metal is deposited prior to the exposure to the arc. This cushioning effect of the molten metal minimizes the digging effect of the arc leading to a reduction in the penetration level. When the electrode is kept normal to the base metal, then intermediate results are obtained. In this case, the heating by the arc and the deposition of the metal take place simultaneously. These welding techniques are shown in Figure 2.3.

One more technique which is employed by the welder is weaving of the arc in the lateral direction to cover a wider area compared to the arc diameter. This is done by the manual welder to improve the appearance of the weld bead. The extent of weaving is usually limited to two or three times the diameter of the electrode. The heat

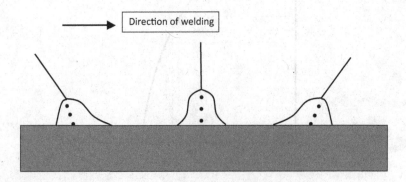

FIGURE 2.3   Forehand, normal, and backhand welding techniques.

input in the base metal is higher when weaving is employed since weaving causes a reduction of the welding speed. When the arc is not weaved at all, the technique is called stringer bead technique and the weld width is just equal to the arc diameter.

Some of the popular arc welding processes and their special characteristics are illustrated below.

2.5 SHIELDED METAL ARC WELDING PROCESS

Shielded metal arc welding is a widely used welding process which is performed manually by a welder. The process employs a coated electrode in the form of a rod which gets consumed during the process. The welder holds the electrode at one end and strikes the arc at the other end of the electrode by touching the base metal with the electrode and retracting it. The electric current is transferred to the electrode near the end where the welder holds the electrode. As the arc consumes the electrode, the length of the electrode steadily comes down and the welder has to feed the electrode in order to maintain the correct arc length. When the length of the electrode is insufficient to proceed with welding, then the short bit of remaining electrode called the **electrode stub** is discarded and a fresh electrode is taken. Thus, the welding process is not continuous but intermittent in nature. Figure 2.4 shows the schematic view of the shielded metal arc welding process.

The protection for the arc from the atmospheric gases is provided by the coating of the electrode, which contains a chemical substance known as flux. The flux coating burns under the arc heat and generates gases which shield the arc from atmospheric gases like nitrogen or oxygen. Apart from forming protective gases, the flux also forms a liquid slag which mixes thoroughly with the liquid weld metal. The slag can chemically react with the liquid metal, and by controlling the chemical reactions between the slag and metal, the metallurgical contaminants in the weld metal may be effectively removed. The removal of the contaminants will result in cleaner metal which has superior mechanical properties. After the arc moves away, the liquid slag which has a lower density floats to the top of the liquid metal and forms a

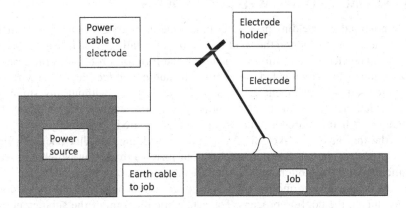

FIGURE 2.4 Schematic diagram of the shielded metal arc welding process.

protective layer over the deposited weld metal. This gives additional protection for the hot metal during the cooling phase. After complete cooling, the slag layer has to be removed thoroughly.

Based on the nature of the flux coating, the electrodes are classified as cellulose coated electrodes, rutile coated electrodes and basic coated electrodes. The coating has a major influence on the running characteristics of the electrode, heat generated in the arc column, the shape of the weld metal and the mechanical properties of the deposited weld metal.

The cellulose coated electrodes predominantly contain an organic substance called cellulose, which burns violently in the arc zone and provides essentially a gas cover. Due to the violent burning of the organic matter, the arc produces a deeper penetration in the base metal. The rutile coated electrodes contain rutile as the main constituent, which imparts good arc stability. The running characteristics of the arc are good, but the required cleanliness of the weld metal is not achieved in this case. When cleanliness of the deposited weld metal is needed, then the choice is a basic coated electrode which has a flux coating consisting of a mixture of basic oxides. These oxides participate in the slag metal reaction in the arc zone and remove the contaminants effectively. But the running characteristics of the electrode are not good, and a skilled welder is needed for welding.

The current levels employed in this process are in the range of 50–120 A, and the corresponding voltages are in the range of 10–20 V. Higher current levels will result in excessive heating of the electrode, which may affect the flux coating. The welding speed is in the range of 1–3 mm/s.

The amount of heat which is realized in the base metal is less than the power generated at the arc. Some heat is lost to the atmosphere through radiation and convection losses. In addition, spatter during welding takes away some heat from the arc zone. Hence, the efficiency of process will be in the range of 70%–80% only.

This welding process has several advantages like positional welding capabilities, cheaper equipment, availability of electrodes for a variety of base metals, etc., but has a serious limitation of lower productivity due to the intermittent operation.

2.6 SUBMERGED ARC WELDING PROCESS

The submerged arc welding process is an automatic welding process in which the welding arc is moved along the weld line by an automatic device like a motorized carriage. The electrode in this case is not in the form of a rod but in the form of a continuous wire which is fed uniformly. The automated feeding of the wire and the motorized arc movement ensure continuous welding without any stoppage. The protection for the arc from the atmosphere is provided by granulated flux particles which are poured around the arc so that the arc is completely submerged under the flux layer. Unlike shielded metal arc welding, the flux does not form gases to protect the arc from atmosphere but forms a thick slag which can react chemically with the molten metal. After the arc moves away, the molten slag layer floats to the top of the molten pool and covers the deposited weld metal completely during the cooling process. The schematic diagram of the submerged arc welding process is shown in Figure 2.5.

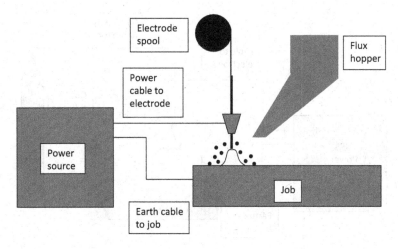

FIGURE 2.5 Schematic diagram of the submerged arc welding process.

Due to the blanketing action of the flux layer, the arc is highly restrained even at high current levels, and there is no violent behaviour of the arc. Hence, the spatter losses are nearly eliminated in the process. The convection and radiation heat losses to the atmosphere will also be a minimum, and hence, the process efficiency is highest among all arc welding processes and will be in the range of 90%–98%. The current levels employed are also highest and will be in the range of 300–600 A with the corresponding arc voltage being in the range of 30–40 V. Because of the high heat input, high welding speeds in the range of 3–10 mm/s can be employed.

The mechanical properties of the deposited weld metal will be decided by the slag metal reactions, and use of basic fluxes gives good mechanical properties, especially notch toughness.

The major advantage of the process is the high deposition rates and absence of radiation from the arc. The major disadvantage of the process is that it can be carried out only in downhand position.

2.7 GAS TUNGSTEN ARC WELDING PROCESS

The gas tungsten arc welding process employs a non-consumable electrode made of tungsten, and the arc is struck between the tungsten electrode and the base metal. The heat which is liberated near the non-consumable electrode has to be removed so that the electrode does not get overheated during the process. This is done by external gas cooling or water cooling of the welding torch which houses the electrode. The protection for the arc zone from the atmosphere is provided by an inert gas such as argon or helium which is made to flow around the arc uniformly. Welding may be performed with or without the additional filler metal in the form of rod or wire. The welding process may be either manual or automatic. When the welding is performed manually, the welder will have the welding torch in one hand and the filler

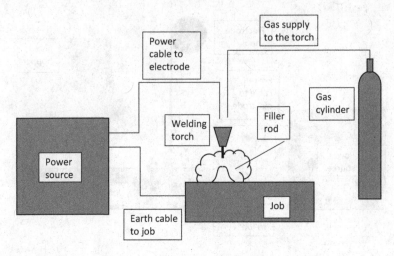

FIGURE 2.6 Schematic diagram of the gas tungsten arc welding process.

metal in the other hand. The arc is moved slowly along the weld line, and the filler metal is added to the molten metal pool as and when required. Figure 2.6 shows the schematic diagram of the gas tungsten arc welding process.

A high level of cleanliness of the arc zone is maintained in the process, and there is no possibility of any contaminant entering the molten metal zone. Hence, the welding process produces very clean weld metal which has good mechanical properties. The process is highly amenable for root pass welding of pipes, tubes and thin sheets. The appearance of the weld bead is dependent on the skill of the operator.

The type of power source connection to the electrode and base metal has a great effect on the pattern of heat liberation in the arc zone. This variable is termed as polarity. When the electrode is connected to the negative terminal of the DC power source and the base metal is connected to the positive terminal, maximum heat is liberated near the base metal and consequently good penetration is obtained in the base metal. This type of connection is called straight polarity.

When the electrode is connected to the positive terminal of the DC power source and base metal is connected to the negative terminal, the type of connection is known as reverse polarity. In this case, maximum heat is liberated near the electrode and much less heat is liberated near the base metal. When AC power is employed, equal amount of heat is liberated near the base metal and the electrode.

The process efficiency is the lowest among arc welding processes as the heat liberated near the electrode is completely lost to the external gas or water cooling arrangement. The process efficiency depends on the polarity employed. While the efficiency can be assumed to be in the range of 45%–60% for DC straight polarity, it is much lower at 25%–30% for the case of DC reverse polarity.

The heat input is the lowest in the case of gas tungsten arc welding process; the current is in the range of 20–150 A, and voltage is in the range of 10–20 V. The welding speed also will be lower at 0.5–3 mm/s.

The power source employed is of constant current type even for automatic welding applications as any increase in current level may lead to melting of the tungsten electrode. To achieve uniform appearance of the weld bead in automatic welding, an automatic voltage correction unit is usually employed. The unit measures the arc voltage and compares it with the set voltage. If the arc voltage is higher than the set voltage, then the electrode is moved towards the base metal. On the other hand, if the measured arc voltage is less than set voltage, then the electrode is moved away from the base metal. Through this control, the arc length is maintained during automatic welding.

The main advantage of the process is the good mechanical properties of the deposited metal and the suitability for root pass welding. The major disadvantage of the process is the lower deposition rate and high level of welder skill.

2.8 GAS METAL ARC WELDING PROCESS

In the gas metal arc welding process, the electric arc is struck between the base metal and a metal electrode in the form of a wire which is supplied continuously by a motorized drive. The protection for the welding arc is provided by the use of a shielding gas which may be inert or active. The shielding gas has an influence on the metal transfer and the shape of the weld bead. The movement of the welding head along the weld line may be manual or automatic. The schematic diagram of the gas metal arc welding process is shown in Figure 2.7.

The electrode in the gas metal arc welding process may be a solid wire or a tubular wire. In the case of tubular wire, flux is packed in the core of the wire and hence, the process is called a flux cored arc welding process. The addition of flux gives many advantages to the process like stable arc, smooth metal transfer, good finish of the deposited metal, etc.

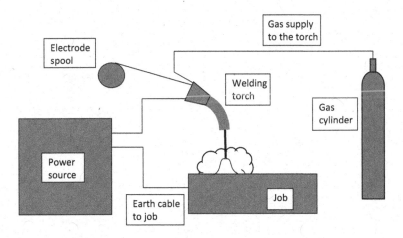

FIGURE 2.7 Schematic diagram of the gas metal arc welding process.

Unlike submerged arc welding, the process can be employed in different positions. Being an open arc welding process, the currents employed are in the range of 80–200 A and the voltage is in the range of 20–30 V. The welding speed will be in the range of 2–5 mm/s. The process efficiency lies between 70% and 80%.

The main advantage of the process is the ability to weld continuously in different positions. The main disadvantage of the process is the high cost of the welding system.

3 Thermal Cycles and Heat Flow in Welding

The unique feature of arc welding processes is the localized heat input which sets up complex temperature profiles in the base metal. While localized heating is desirable for achieving coalescence of the base metal, it has some aftereffects. Hence, the thermal history of the base metal which is the root cause of these metallurgical and mechanical phenomena must be fully understood for controlling the detrimental aftereffects within limits. The thermal history is influenced by many parameters, and unless the various issues are considered collectively, it is difficult to comprehend the behaviour of the welded component.

3.1 HEATING AND COOLING CYCLES

The heat energy which is transferred to the base metal by the electric arc can be calculated as the product of process efficiency, welding current and arc voltage. This arc heat is transferred to a small area in the base metal which lies underneath the arc. Since the arc is continuously moving, the heat which is transferred at a particular point depends on the time taken by the arc to cross the point. The intense arc heat raises the temperature of the metal almost instantaneously. The heat flux and the material properties, especially the specific heat, determine the extent of superheat beyond the melting point and in some instances there may even be some vaporization of the metal in this zone. As the heating is intense, the high temperatures are experienced only by points which are directly underneath the arc, and points outside this area will still be at lower temperature.

But as the arc moves away, the temperature of the weld zone starts falling due to dissipation of heat from the molten zone. The heat is mainly dissipated through conduction process to the surrounding cooler regions, and there will also be convection and radiation heat loss to the environment. The heat loss through conduction process depends on the thermal conductivity of the material, the thickness of the base metal and the temperature gradient. The conduction heat transfer is incremental in nature. The temperature difference leads to flow of heat, and as the heat flows towards a point, the temperature of the point rises. This in turn causes further heat conduction to the next cooler point. In this way, the heat is transferred up to the farthest point in the base metal.

The temperature of a point in the base metal rises as long as the heat flow towards the point is more than the heat flow away from it. When the heat flow towards the point becomes less, the temperature starts falling. Thus the various points in the base metal experience a heating cycle and a cooling cycle; correspondingly the temperature rises to a peak value, and thereafter the temperature falls down. The peak temperature and the rate of heating are dependent on the distance of the point from the weld. Since the conduction heat flow requires time, the time taken by a point to reach the peak temperature increases with distance from the weld.

The thermal gradient is very steep initially but as the conduction heat flow takes place, the temperature difference gradually comes down. Ultimately, the temperatures in the base metal get equalized after a long time. The time taken for equalization of temperatures depends on the thermal conductivity of the material.

3.2   HEAT FLOW IN BASE METAL

One of the parameters which has a major influence on the heat conduction process is the thickness of the base metal. As the thickness of the base metal increases, the pattern of heat flow changes from two-dimensional to three-dimensional. The abstraction of heat in the case of a thick plate is much faster due to the large mass of surrounding metal which acts as a heat sink. There are multiple paths for the heat flow in a thicker metal unlike a thin plate as shown in Figure 3.1. It is difficult to clearly demarcate the thickness above which the heat flow can be termed as three-dimensional in nature. In general, the heat flow is strictly three-dimensional in all thicknesses, but for simplification purposes, the flow can be assumed to be two-dimensional in some special cases. Good judgment is needed to make the simplifying assumption.

The heat losses through convection and radiation depend on the surface area of the base metal which is exposed to the environment and the temperature difference between the base metal and environment. The ratio of the surface area to the total volume of metal is greater when the plate thickness is less. Hence, the heat loss by convection and radiation become very significant in the case of thin plates compared to thick plates. The convection heat loss depends on the convection coefficient between the metal surface and the environment. The coefficient depends on the nature of cooling viz. forced cooling and natural cooling. Any external water or gas cooling of the plate will result in higher heat loss through convection. The radiation heat loss depends on the emissivity of the surface which in turn depends on the surface finish of the base metal. The selection of the heat loss parameters must be made carefully to reflect the actual welding conditions.

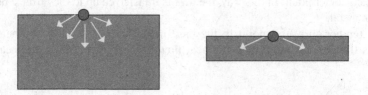

FIGURE 3.1   Three-dimensional and two-dimensional heat flow in the welded plate.

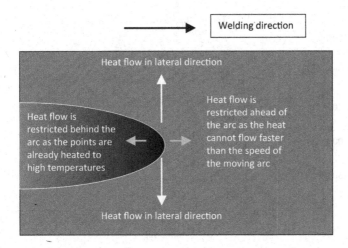

FIGURE 3.2 Heat flow in a welded plate with continuously moving arc.

In the case of a stationary heat source, the heat flow will be equal in all directions, but in the case of a continuously moving heat source, there are limitations to the direction of heat flow as shown in Figure 3.2. In a plate where the arc is moving forward continuously, the heat flow takes place mainly in the lateral direction as shown in the figure. The heat flow is restricted in the backward direction as this region has already been heated by the arc at the previous instance. The heat flow is also restricted in the forward direction as the speed of welding is higher than the rate of conduction of the heat. Thus most of the heat flow takes place in the transverse plane with reference to the arc movement direction. This fact is usually employed for simplifying the complex three-dimensional analysis into a simpler two-dimensional analysis. In this case, the plane of interest is the weld cross section and the arc movement is normal to this plane. This cross sectional model is applicable when the plane of interest lies in the middle of a long weld and the welding arc moves at a uniform rate.

3.3 GRAPHICAL PLOTTING OF RESULTS

The typical temperature versus time graph for various points in the weldment is as shown in Figure 3.3. The thermal cycle is a useful method of analysis of thermal history in welding. The advantage of this method of representation is that the effect of welding on different points can be studied and the rate of heating, the peak temperatures reached, the time for which the point remains at high temperature and the rate of cooling are determined for various points in the weldment. This information is useful for the assessment of metallurgical changes in the base metal and the residual stress build up. But the limitation of this method of representation is that the analysis is applicable for a given point and hence the spatial distribution of heat in the base metal is not captured here.

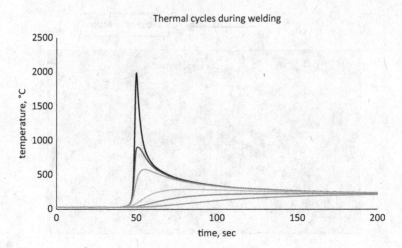

FIGURE 3.3 Typical thermal cycle plot of a welded plate. The different lines represent the thermal cycle of points at varying distances from the weld line.

The other form of representation of the heat flow is the plotting of isotherms on the weld surface. In this method, various isotherms for a given time are plotted on the top plane of the component. The isotherms are lines which connect all the points which are at the same temperature at the given instance. Thus an isotherm corresponding to the melting point of the metal gives an idea of the molten pool shape at the given time interval. The isotherms corresponding to higher temperatures are found to encircle the molten pool shape closely while the isotherms corresponding to lower temperatures are found to spread out in the component. A typical representation of isotherms is shown in Figure 3.4. The isotherms are closely bunched ahead of a moving arc, whereas the isotherms behind the arc

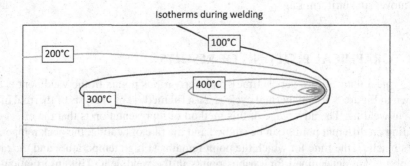

FIGURE 3.4 A typical plot of isotherms in a welded plate. The different lines represent different temperatures.

are well spaced out. The flow of heat in the lateral direction is also revealed by the isotherms. The heat is initially concentrated near the arc and gets dissipated after some time.

The isotherms give a good picture of the spread of heat in the base metal for various instances but do not represent the rate of heating and cooling of these points. The isotherms are usually drawn for the top surface of the plate and can also be drawn for different layers of metal in the thickness direction. In the case of thin plates, the isotherms on the top surface and the isotherms at the bottom surface are nearly the same as the heat flow in the thickness direction is limited. The temperature drop between the points on the top surface and bottom surface will be small in this case.

This fact that the isotherms are nearly identical for the top and bottom surfaces in the case of a thin plate is used for simplifying the complex three-dimensional analysis into a two-dimensional analysis. Here the two-dimensional plane of interest is the midplane of the plate, and the heat flow in the thickness direction is neglected.

In addition to these methods of representing the thermal history, the temperature distribution in longitudinal direction and transverse direction are also plotted for various time intervals. The temperature is plotted with distance in the longitudinal direction, and a typical plot is shown in Figure 3.5. The location of the arc at the given instance is indicated by the high temperature region in the graph. The temperature gradient is steep ahead of the arc indicating that the heat flow is restricted, whereas the temperature gradient behind the arc is less steep indicating cooling of the weld metal. Likewise the plot of temperature versus transverse distance is as shown in Figure 3.6. The temperatures are highest at the weld zone and the temperature gradient is quite steep in the transverse direction. These plots are drawn for a given instance of time, and a series of such plots for various time instances are required to understand the heat flow in the base metal.

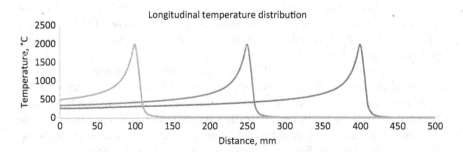

FIGURE 3.5 Typical longitudinal temperature distribution in a welded plate. The different lines represent the temperature distribution corresponding to different time intervals.

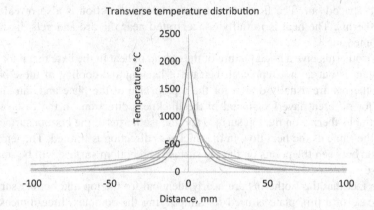

FIGURE 3.6 Typical transverse temperature distribution in a welded plate. The different lines represent the temperature distribution corresponding to different time intervals.

3.4 FACTORS INFLUENCING THERMAL CYCLES

There are physical phenomena which have some influence on the thermal analysis, and proper incorporation of these phenomena in the model is necessary for getting accurate results. The material properties like specific heat and thermal conductivity do not remain constant but vary with temperature. This temperature dependency of the material properties has a major influence on the accuracy of the temperature results. Accounting this effect makes the analysis non-linear in nature, and the consequent temperature history can be determined only with the help of an iterative procedure. It must be borne in mind that the reliability of the thermal properties is not very good especially at high temperatures. The values are sometimes extrapolated as the experimental determination is difficult at high temperatures.

Another major factor which plays an important role in the heat transfer from the molten pool is the strong convection currents in the molten pool. The molten pool experiences severe stirring due to the action of magnetic forces, and the resulting convection current causes high heat loss to the surrounding regions. This heat loss will be experienced only when the metal remains in the molten condition. To account for this heat loss, the thermal conductivity is set to a high value at temperatures beyond the melting point of the material.

The distributed arc heat is another major factor which influences the accuracy of the thermal results. When the arc is assumed to spread over a larger area, the heat density is reduced and correspondingly the temperature of the molten puddle is lower. If the area under the arc is less, then it may lead to high temperature in the molten metal and there may even be vaporization of the metal. The shape of the arc is also very important and it has to be selected carefully for close results.

The latent heat is another phenomenon which has a minor influence on the thermal cycles. The latent heat is the amount of heat which is absorbed by the metal when it reaches the melting temperature during the heating cycle. The same latent heat is subsequently released by the metal during the cooling cycle. Since this absorption and liberation of heat take place within a short duration, it does not have a major effect on the accuracy of thermal history of the plate.

3.5 SIMPLIFICATION OF THREE-DIMENSIONAL MODEL

As mentioned, the three-dimensional analysis is applicable for all the components, but the analysis is expensive to perform and time consuming. In cases where some simplifying assumptions will not affect the accuracy of the results, a two-dimensional analysis may be performed. However, the assumption should be made carefully after evaluating the suitability for the particular condition.

One method of simplification is to perform the thermal analysis for a typical cross section which lies perpendicular to the direction of movement of the arc. The analysis for one typical plane will be applicable for all other planes in the component provided the time lag is suitably considered. The time lag between the thermal cycles in different planes is given as the distance between the planes divided by the speed of welding. The plane of interest receives the heat during the time when the arc is crossing the plane, and there will be no heat input in the plane prior to or after this arc movement.

The assumption that the heat flow is restricted to the transverse direction is not entirely valid especially for points close to the molten zone. The cross sectional model puts a restriction on the longitudinal flow of heat, and consequently the temperatures predicted in the molten pool using this model will be on the higher side. In addition, the cross sectional model is not valid for short welds. The demarcation between the long weld and short weld is not clearly defined, and the researcher has to judge the suitability of this model for a particular condition.

Another simplifying assumption for running a two-dimensional analysis is to neglect the heat flow in the thickness direction of the plate. This is true in the case of thin plates where the temperature gradient between the top and bottom surfaces is small. The limiting thickness up to which the assumption can be applied is to be decided carefully by the researcher. This model gives good results for points well away from the molten zone. But in the weld zone, there will be a steep thermal gradient, which is ignored by this model.

4 Finite Element Analysis

The finite element method is a numerical technique for the analysis of engineering components. In this method, the operating domain is divided into a number of sub-domains called elements. The elements are connected to each other by the corner points which are called nodes. The field variable is assigned to the nodes and within each element, the field variable is assumed to vary as per linear or quadratic relationship while maintaining continuity at the nodes. The field variable is thus said to be piecewise continuous and is expressed in terms of the interpolation function or shape function for the elements and the nodal values of the field variable.

Using a minimization process for the functional formulation of the governing equation, the partial differential equation which represents the three-dimensional transient heat flow is converted into a set of linear equations which are best represented in matrix form. The equation is evaluated for each element, and all these elemental matrices are finally assembled to form a global matrix. Then the particular conditions known as boundary conditions are imposed and the matrix equation is solved to obtain the nodal field variable. In a transient analysis, the procedure is repeated for a series of time steps in order to get the nodal values of field variable for various time intervals.

4.1 SHAPE FUNCTION

In the case of thermal analysis, the field variable is the temperature, and the temperature at any point is expressed in terms of shape functions and nodal temperatures as

$$T = [N]. \{T\} \tag{4.1}$$

The evaluation of shape function begins with the assumption of a polynomial relationship for representing the temperature as a function of spatial coordinates. The evaluation of shape functions in the case of a simple one-dimensional analysis and a two-dimensional analysis using a quadrilateral element are presented in the following sections.

4.1.1 SHAPE FUNCTION FOR A ONE-DIMENSIONAL SIMPLEX ELEMENT

In the case of a one-dimensional analysis, the simplex element is a line element with two nodes at the two ends as shown in Figure 4.1. Let the x coordinate of the two nodes be x1 and x2. Let the temperature at the two nodes be T1 and T2. The temperature at any point within the element is represented by the following linear relation.

$$T = a0 + a1. x \tag{4.2}$$

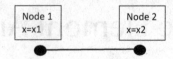

FIGURE 4.1 One-dimensional simplex element.

The equation contains two unknown parameters a0 and a1 which may be evaluated using the known values of temperatures at the two nodes.

Substituting the x coordinate values and the nodal temperatures, we get

$$T1 = a0 + a1. x1$$

$$T2 = a0 + a1. x2$$

By solving for the two unknown parameters from the two equations, we get

$$a0 = T1 - (T2 - T1) / (x2 - x1). x1$$

$$a1 = (T2 - T1) / (x2 - x1)$$

Substituting the values of a0 and a1 in these equations, we get

$$T = T1 - (T2 - T1) / (x2 - x1). x1 + (T2 - T1) / (x2 - x1). x$$

Simplifying,

$$T = ((x2 - x) / (x2 - x1)). T1 + ((x - x1) / (x2 - x1)). T2 \qquad (4.3)$$

This can be written as T = N1. T1 + N2. T2, where N1 and N2 are the shape functions for the line element.

$$N1 = (x2 - x) / (x2 - x1)$$

$$N2 = (x - x1) / (x2 - x1)$$

In matrix form, the equation can be written as

$$T = \begin{bmatrix} N1 & N2 \end{bmatrix} \begin{Bmatrix} T1 \\ T2 \end{Bmatrix} \qquad (4.4)$$

The shape function of a node represents the influence exerted by the node in determining the field variable within the element. The combined influence of the nodes in an element, which is the sum of all the shape functions, is always equal to unity at any point within the element. The shape function for a node decreases from a value of unity at the node location to a value of zero in the other node locations.

4.1.2 Shape Function for a Four Noded Quadrilateral Element

In the case of a four noded quadrilateral element with four nodes as shown in the figure, the field variable can be expressed as

$$T = a0 + a1.\, x + a2.\, y + a3.\, x.y \qquad (4.5)$$

The above polynomial function contains linear terms and one higher order term. The four unknown constants a0, a1, a2 and a3 in the equation can be determined from the four nodal temperatures and the nodal coordinates.

In order to simplify the analysis, the origin of the coordinate system is assumed to be located at node 1. The element has a length equal to L and width equal to W. The coordinates of the four nodes are shown in Figure 4.2.

Substituting the four values of x and y, we get four equations, and by solving the equations, the values a0, a1, a2 and a3 can be obtained.

$$a0 = T1$$

$$a1 = (T2 - T1)/L$$

$$a2 = (T4 - T1)/W$$

$$a3 = (T1 - T2 + T3 - T4)/L.W$$

So the interpolation function for temperature at any point x and y can be written as

$$T = T1 + (T2 - T1).\,(x/L) + (T4 - T1).\,(y/W)$$
$$+ (T1 - T2 + T3 - T4).\,(x/L).\,(y/W)$$

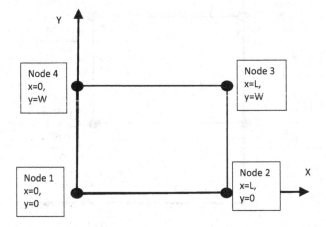

FIGURE 4.2 Four-noded quadrilateral element. The origin of the coordinate system is at node 1.

Simplifying the equation, we get

$$T = T1.(1 - x/L - y/W + x/L. \, y/W) + T2.(x/L - x/L. \, y/W)$$
$$+ T3. \, (x/L . \, y/W) + T4. \, (y/W - x/L. \, y/W)$$

(4.6)

The expression can also be written as

$$T = N1. \, T1 + N2. \, T2 + N3. \, T3 + N4. \, T4$$

(4.7)

where:

$$N1 = (1 - x/L - y/W + x/L. \, y/W) = (1 - x/L). \, (1 - y/W)$$

$$N2 = (x/L - x/L. \, y/W) = x/L. \, (1 - y/W)$$

$$N3 = (x/L . \, y/W)$$

$$N4 = (y/W - x/L. \, y/W) = y/W. \, (1 - x/L)$$

Alternately, if the coordinate system is selected as shown in Figure 4.3, the shape functions are more general in this case. The origin of the coordinate system is not at node 1 in this case. The coordinates at the four nodes are (0, W1), (L, W1), (L, W2) and (0, W2), respectively.

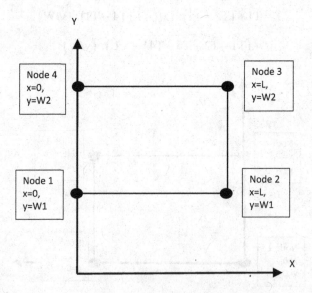

FIGURE 4.3 Two-dimensional rectangular element. The origin of the coordinate system does not coincide with any node.

In this case, the shape functions are determined as

$$N1 = (1 - x/L) \ (W2 - y)/(W2 - W1)$$

$$N2 = (x/L) \ (W2 - y)/(W2 - W1)$$

$$N3 = (x/L) \ (y - W1)/(W2 - W1)$$

$$N4 = (1 - x/L) \ (y - W1)/(W2 - W1)$$

4.2 FORMULATION OF EQUATION

The shape functions are a function of geometry, whereas the nodal temperatures are independent of geometry. Hence, the first derivative of the temperature with respect to x can be expressed as

$$dT/dx = [\ dN/dx \ ] . \{ \ T \ \}$$

Expressing the derivatives with respect to x, y and z in matrix form,

$$
\begin{Bmatrix} dT/dx \\ dT/dy \\ dT/dz \end{Bmatrix} =
\begin{bmatrix} dN1/dx & dN2/dx & dN3/dx & \ldots \\ dN1/dy & dN2/dy & dN3/dy & \ldots \\ dN1/dz & dN2/dz & dN3/dz & \ldots \end{bmatrix} \times
\begin{Bmatrix} T1 \\ T2 \\ T3 \\ .. \\ .. \end{Bmatrix}
\tag{4.8}
$$

The governing equation for the transient thermal analysis is the partial differential equation as follows.

$$\partial/\partial x \ (kx.\partial T/\partial x) + \partial/\partial y \ (ky.\partial T/\partial y) + \partial/\partial z \ (kz.\partial T/\partial z) = \rho c.\partial T/\partial t \tag{4.9}$$

where kx, ky and kz are the thermal conductivities of the material in x, y and z directions, ρ is the density of the material and c is the specific heat of the material.

Within the operating domain, there are some nodes which receive the external heat flux q along a surface. Similarly, there are nodes lying at the free surface from which the heat is lost by convection. The convection heat loss is expressed as h. $(T - T\infty)$, where $T\infty$ is the ambient temperature and h is the convection coefficient.

Incorporating the boundary conditions in the above equation, we get the equation as

$$\partial/\partial x. (kx.\partial T/\partial x) + \partial/\partial y. (ky.\partial T/\partial y) + \partial/\partial z. (kz.\partial T/\partial z)$$
$$+ h. (T - T\infty) + q = \rho c.\partial T/\partial t \qquad (4.10)$$

Using a minimization function for the functional formulation of the differential equation, the above partial differential equation is represented in matrix form as

$$[c] . \{dT/dt\} + [k] \{ T \} = \{ f \} \qquad (4.11)$$

where

[c] is the heat capacitance matrix written as $\int \rho c. [N]^T. [N]. dV$,

[k] is the heat conductance matrix written as $\int [B]^T. [D]. [B] dV + \int h. [N]^T. [N]. dS$,

{f} is the force matrix written as $\int q. [N]^T dS + \int h. [N]^T. T\infty. dS$, and

[B] is the matrix containing derivatives of shape functions and is written as follows.

$$[B] = \begin{bmatrix} dN1/dx & dN2/dx & dN3/dx & \cdots \\ dN1/dy & dN2/dy & dN3/dy & \cdots \\ dN1/dz & dN2/dz & dN3/dz & \cdots \end{bmatrix}$$

[D] is the matrix containing thermal conductivities written as

$$[D] = \begin{bmatrix} kx & 0 & 0 \\ 0 & ky & 0 \\ 0 & 0 & kz \end{bmatrix}$$

The elemental matrices [c] and [k] are assembled for all the elements and global [C] and [K] matrices are formed. The equation is then solved to get the nodal temperatures {T} for various time intervals. For solving the transient analysis, the term $\partial T/\partial t$, T and F are written as

$$\partial T/\partial t = (T2 - T1)/\Delta t$$

$$T = (T2 + T1)/2$$

$$F = (F2 + F1)/2$$

Where T1 and T2 are the temperature of the previous step and current time step respectively, F1 and F2 are the thermal load corresponding to the previous time step and current time step respectively and Δt is the time increment. Thus the matrix form of the equation is written as,

$$[C] \cdot (T2 - T1)/\Delta t + [K] \cdot$$

$$(T2 + T1)/2 = (F2 + F1)/2$$

Simplifying,

$$[[C]/\Delta t + [K]/2] \cdot \{T2\}$$
$$= \{(F2 + F1)/2\} + [[C]/\Delta t - [K]/2] \cdot \{T1\} \tag{4.12}$$

This simplified equation can be expressed as

$$[A] \cdot \{T2\} = \{B\}$$

Thus, knowing the values of [C], [K], {F1}, {F2} and {T1} matrices, the unknown matrix {T2} can be evaluated. Repeating the procedure, the nodal temperature values can be determined for various time steps from the start to the end. This is known as time marching scheme.

The transient thermal analysis is non-linear in nature since the material properties such as thermal conductivity and specific heat are temperature dependent. Thus the equation becomes

$$[C(T)] \{dT/dt\} + [K(T)] \{T\} = \{F\} \tag{4.13}$$

where the matrices [C] and [K] are both temperature dependent. The equations cannot be solved in a single step, but an iterative technique is needed to get the temperature results. The iteration must be repeated until convergence is achieved. Newton Raphson's method is the usually adopted technique for such nonlinear problems. In this technique, the matrices are iteratively evaluated using the previous set of temperature results, and the error in accounting the force terms is determined. The procedure is repeated until the error due to unaccounted force is a small fraction of the initial force.

4.3   FINITE ELEMENT ANALYSIS USING ANSYS

ANSYS is a general purpose finite element software which can be successfully employed for the transient thermal analysis of welding problems. The software consists of three modules such as preprocessor, solution, and post-processor. In the preprocessor module, the element type and the material properties are selected and the meshing of the domain of interest is carried out. In the solution module, the time interval for transient analysis, the convection heat loss and arc heat input are entered and the problem is solved for a given time step. If the analysis is of transient type, then the analysis is repeated for different time steps. In the post-processor module, the temperature results for various sets can be viewed. In addition, the animation feature enables one to visually picturize the flow of heat in the domain of interest.

The selection of element type is an important step, and the researcher has to decide whether the analysis is two-dimensional or three-dimensional in nature. ANSYS has a big library of elements, and either a lower order element or a higher order element can be selected. In some problems, multiple types of elements may be selected in the same problem. The material properties of interest for the transient thermal analysis are thermal conductivity, specific heat, and density. These properties can be input as a function of temperature.

Once an element type is selected and the material properties are entered, then the meshing of the domain of interest is performed. In ANSYS, the meshing of the region of interest can be done in two ways such as by direct generation of nodes and elements or by solid modelling with automatic mesh generation within the domain. The direct generation is a simpler method, and in this method, the nodes and their coordinates are defined first. Using the "ngen" command, multiple sets of nodes may be generated from a master set. The elements are defined subsequently from the nodes. Smaller element size is selected for regions close to the weld where steep temperature gradients are expected and larger element size is selected for far off points. Multiple elements may be generated from a master set using the "egen" command. This procedure is easy to apply, but the disadvantage of this procedure is that there will be elements with unfavourable shapes.

In the alternate method, which is called solid modelling, the meshing is performed by dividing the domain into various volumes or areas. The user has to generate the desired areas or volumes using keypoints and lines. Then the size of the desired mesh near each keypoint is entered using the "kesize" command. The software automatically generates the nodes and elements which satisfy the user's requirement of mesh spacing near each keypoint. The disadvantage with this procedure is that the node numbers which are generated are not uniform. This becomes a problem when the in-plane analysis has to be performed using ANSYS Parametric Design language (APDL).

The meshing of the domain is an important step, and the user may have to repeat it several times until satisfactory results are obtained. In the mesh generation through areas and volumes, the division of the domain into different areas and volumes must be attempted to get the required results.

There is no hard and fast rule in deciding whether a given mesh is satisfactory or not, but the following set of rules must be borne in mind.

1. The elements should not be distorted too much from the ideal shape. For example, if a triangle is selected as the element type, then an ideal shape of the triangle is an equilateral one with the corner angles of 60°. But getting such a mesh is not always possible, and it may be okay if any of the corner angle goes up to 90°. But if an element has a corner angle more than 90°, then the shape is not acceptable.
2. The mesh size should increase gradually along any line. If the increase of mesh size is abrupt along any line, then it is not considered to be a good meshing.
3. The shape factor of the element is important. For a four noded quadrilateral element, the ideal shape will be a square with the length being equal to the width. A selection of rectangle with length being 2–3 times the width is okay, but if the length is several times compared to the width, then such a shape is not okay.

When the meshing is done, ANSYS gives out warnings if the procedure has resulted in the generation of elements with any of the above violations. However, if the researcher chooses to ignore the warnings, then the analysis will be performed anyway. It is left to the researcher to estimate the inaccuracy of such a case.

In the solution module, the researcher has to define the heat losses from the surface nodes. Likewise, the nodes which receive the arc heat at any time interval and the exact values of the heat input must be entered.

The transient analysis has to be performed for several time steps. Instead of manually executing the commands, which is prone for errors, the analysis may be performed using a program which is written in APDL language. The APDL enables a user to perform multiple analyses using a *do* command.

The results of the analysis can be viewed in a general post-processor or time-based post-processor. In a general post-processor, the temperature distribution corresponding to a chosen set is viewed, and in the time-based post-processor, the thermal cycle can be viewed for a selected point. In addition, the post-processor has an option to view the animated results. The progressive heat flow in the plate can be viewed to obtain an overall idea of heat transfer in the component. There is also a provision to view the temperature results along a chosen path such as longitudinal direction or any transverse direction.

5 Arc Heat Model

The efficacy of the finite element model depends to a large extent on the accuracy in the determination of the arc heat input and the arc heat distribution in the base metal. Good judgment is needed in determining the arc shape and size. The initial researchers who employed analytical models assumed the arc to be a point source due to the simplicity of the model, but this assumption leads to a very high temperature in the molten pool. Subsequent researchers tried various models for the arc heat distribution including circular and elliptical shaped models.

In the case of automatic welding processes where the electrode is held perpendicular to the base metal, the area covered by the arc can reasonably be assumed to be a circle. If the electrode is not held perpendicular to the base metal as in the case of forehand and backhand welding techniques, the resulting arc shape may not be a circle but elliptical. When the electrode is tilted towards the forward direction, the arc gets elongated in the forward direction. Similarly, when the electrode is tilted backwards, then the arc may be elongated in the backward direction as shown in Figure 5.1. The heat distribution in the forward and backward regions of the arc may or may not be equal. Hence, the elliptical model of the arc heat has many parameters which have an influence on the temperature results. Unless these parameters are selected correctly, the accuracy of the model will suffer.

The diameter of the arc is another major factor which has a great influence on the temperature results. For a given heat input, a higher arc diameter leads to lower energy density which may be inadequate to cause melting of the base metal surface. If the diameter is smaller, then the higher energy density may lead to overheating of the molten base metal and may even cause vaporization of the metal. The selection of the correct arc diameter is crucial for getting good results.

The arc diameter is directly proportional to the arc length between the electrode tip and the base metal which in turn is proportional to the arc voltage. Figure 5.2 shows the dependence of arc diameter on the arc length. The arc diameter must be selected based on the voltage which is employed during the welding process. In processes like GTAW or SMAW where the arc voltage is in the range of 10–20 V, the arc diameter will be in the range of 4–8 mm. In the GMAW process, the arc voltage will be in the range of 20–30 V, hence it is appropriate to assume the arc diameter to be between 6 and 10 mm. In the SAW process, the voltage will be in the range of 30–40 V and hence, the arc diameter may be selected from the range of 10–16 mm. In the case of open arc processes, the arc diameter may also be judged by viewing the arc during welding.

In addition to the distribution of arc heat along the top surface of the metal, some heat may also be input in the subsurface regions of the base metal. This happens when the force of the arc displaces the liquid metal and the metal lying underneath is directly exposed to the arc. This becomes an important factor in the case of beam welding processes in which the beam may penetrate through the entire thickness of the work piece. The calculation of the arc heat is more complicated in such cases.

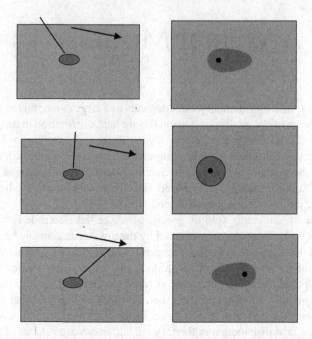

FIGURE 5.1   The shape of the arc in forehand, normal, and backhand welding techniques.

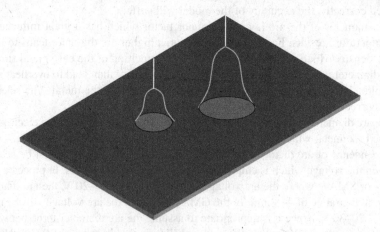

FIGURE 5.2   The diameter of the arc as a function of the arc length.

In the present case, the area covered by the arc is assumed to be a circle, and the arc base is assumed to lie in the top plane of the base metal without displacing the molten metal. The heat flux within the arc region is assumed to vary as per Gaussian distribution, i.e., the heat flux varies exponentially from the centre of the arc to the periphery of the circular zone.

The equation which represents the arc heat flux at any radial distance r from the centre is[1,2]

$$q(r) = 3\eta VI / \pi R^2 . Exp(-3 r^2/R^2) \qquad (5.1)$$

where q (r) is the heat flux in W/mm^2 at a radial distance r from the centre of the circle, R is the radius of the arc circle, η is the welding process efficiency, V is the arc voltage, and I is the arc current.

It is seen from the above equation that the heat flux is the highest at the centre of the circular zone, and this peak value is three times the average value of heat flux. The heat flux drops exponentially from the centre of the arc as shown in Figure 5.3. At the periphery where r = R, the heat flux drops to a value which is only 5% of the value at the centre of the arc. There will be some stray heat flux outside the circle, and this stray heat is not accounted in the model.

If the arc lies in the xy plane and its centre is in the origin of the coordinate system as shown in Figure 5.4, the radial distance r is equal to $\sqrt{(x^2 + y^2)}$ and hence, the arc heat flux can be expressed as

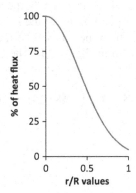

FIGURE 5.3 Gaussian distribution of the arc heat flux as a function of r/R.

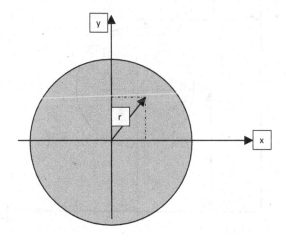

FIGURE 5.4 The arc heat flux when the centre of the arc is at the origin of the coordinate system.

$$q\,(x,y) = 3\eta\,VI/\pi R^2.\,Exp\,\left(-3\left(x^2+y^2\right)/R^2\right) \tag{5.2}$$

The above expression can also be written as

$$q\,(x,y) = 3\eta\,VI/\pi R^2.\,Exp\,\left(-3x^2/R^2\right)\,Exp\,\left(-3y^2/R^2\right) \tag{5.3}$$

If the centre of the arc is at some distance away from the origin of the coordinate system as shown in Figure 5.5, then the arc heat is expressed as

$$q\,(x,y) = 3\eta\,VI/\pi R^2.\,Exp\,\left(-3(x-a)^2/R^2\right).\,Exp\,\left(-3(y-b)^2/R^2\right) \tag{5.4}$$

In this case, a and b are the x and y coordinates of the arc centre as shown in Figure 5.5.

In the case of an arc which is moving at a particular speed v, the arc will be at different locations for different time intervals. If the arc centre was at the origin of the coordinate system corresponding to a time t = 0 and if the arc is assumed to move along the x direction, then the arc will be located at a distance v·t from the origin at any time t as shown in Figure 5.6. The corresponding heat flux can be represented as

$$q\,(x,y,t) = 3\eta\,VI/\pi R^2.\,Exp\,\left(-3(x-vt)^2/R^2\right).\,Exp(-3y^2/R^2) \tag{5.5}$$

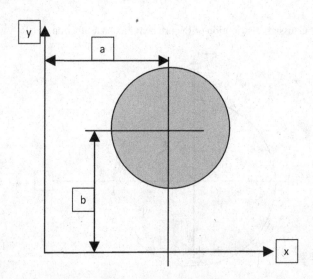

FIGURE 5.5 The arc heat flux when the centre of the arc is at some distance from the origin of the coordinate system.

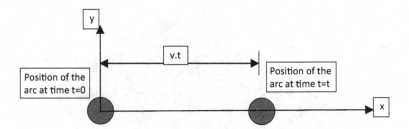

FIGURE 5.6 The position of the moving arc at time t in the coordinate system. The centre of the arc was at the origin of the coordinate system at time t = 0 s.

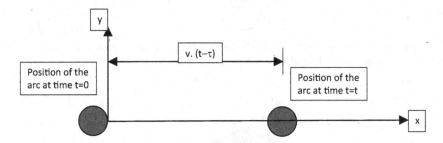

FIGURE 5.7 The position of the moving arc at time t in the coordinate system. The arc was touching the origin of the coordinate system at time t = 0 s.

If the arc was just touching the origin of the coordinate system corresponding to time t = 0 as shown in Figure 5.7, then the arc will be located at a distance v. (t–τ) at a time t where τ is the time lag given by the ratio of arc radius to welding speed R/v. The heat flux in this case can be written as

$$q\,(x,y,t) = 3\eta VI/\pi R^2 . \, \mathrm{Exp}\,(-3(x - v(t - \tau))^2/R^2). \, \mathrm{Exp}(-3y^2/R^2) \qquad (5.6)$$

These relations are useful in determining the arc heat input at any node. The calculation of arc heat for the cross sectional analysis and in-plane analysis are presented below.

5.1 CROSS SECTIONAL ANALYSIS

In a full three-dimensional model as shown in Figure 5.8, the arc lies in a plane which is parallel to the zx plane and moves steadily in the z direction as shown. Initially, at time t = 0, the arc was touching the origin as shown in the figure. In this case, the arc heat liberated at a point represented by the coordinates z and x corresponding to a time t is given by the relation

$$q\,(z,x,t) = 3\eta VI/\pi R^2 . \mathrm{Exp}\,(-3(z - v(t - \tau))^2/R^2). \, \mathrm{Exp}(-3x^2/R^2) \qquad (5.7)$$

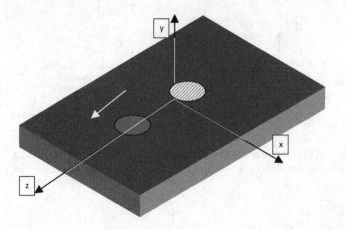

FIGURE 5.8 A full three-dimensional model of the welded plate. The arc moves in a plane which is parallel to zx plane and moves in the z direction.

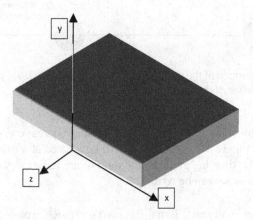

FIGURE 5.9 Typical cross section for analysis.

If a cross section in the xy plane lying at z = 0 is considered as shown in Figure 5.9, then the above equation becomes

$$q\,(x,t) = 3\eta VI/\pi R^2 . \exp\left(-3(v(t-\tau))^2/R^2\right). \mathrm{Exp}(-3x^2/R^2) \tag{5.8}$$

In the above expression, the heat liberated due to the movement of the arc in z direction is represented by the term $\exp\left(-3(v(t-\tau))^2/R^2\right)$. The heat input initially increases and reaches a peak value at time $t = \tau$ and thereafter the heat input again decreases. When the time reaches a value $t = 2.\tau$, then the arc heat once again goes to a value which is nearly zero. The cross section receives the arc heat during the interval between $t = 0$ to $t = 2.\,\tau$ and beyond that time, there will be only a cooling cycle with no arc heat input. These are illustrated in Figures 5.10 through 5.13.

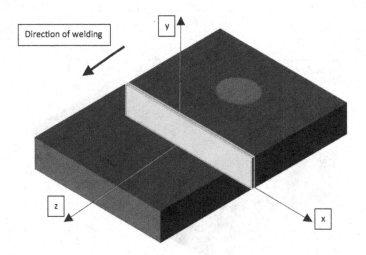

FIGURE 5.10   Movement of the arc during welding process and the plane of interest is shown.

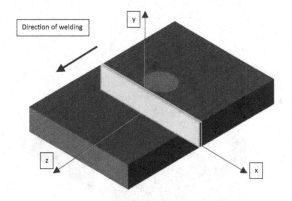

FIGURE 5.11   Analysis corresponding to time t = 0. The arc is just touching the plane of interest. Arc heat input commences from this time step.

In order to carry out the two-dimensional thermal analysis, the cross section of the plate which lies in the xy plane is taken into consideration. By invoking the condition of symmetry, only one half of the plate is considered for the analysis as shown in Figure 5.14.

To calculate the arc heat input for the cross sectional analysis, the procedure is as follows. First, the two-dimensional domain of interest is created and after the mesh generation, the nodes which are covered by the arc are identified. In this case, the arc heat is experienced along a line on the top surface whose length is equal to the radius of the arc as shown in Figure 5.15. It is a good practice to divide the line

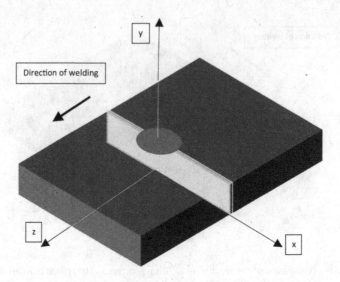

FIGURE 5.12 Analysis corresponding to time t = τ. The centre of the arc is directly over the plane of interest. Heat input is maximum for this time step.

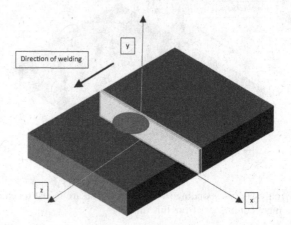

FIGURE 5.13 Analysis corresponding to time t = 2τ. Arc is touching the plane of interest. Heat input ends at this time step.

corresponding to the arc radius into five or more elements. The arc heat has to be calculated for the corresponding nodes.

From the FEM analysis, the nodal values of heat input along a line which receives the heat flux q are determined by performing the following integration function.

$$\{ F \} = \int q. \left[N \right]^{T} dx \tag{5.9}$$

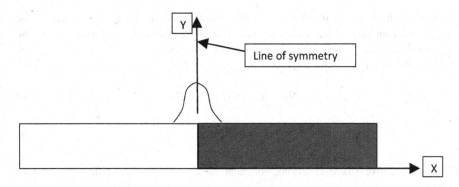

FIGURE 5.14 Analysis of a plane taking the symmetry into consideration. Arc moves perpendicular to the plane.

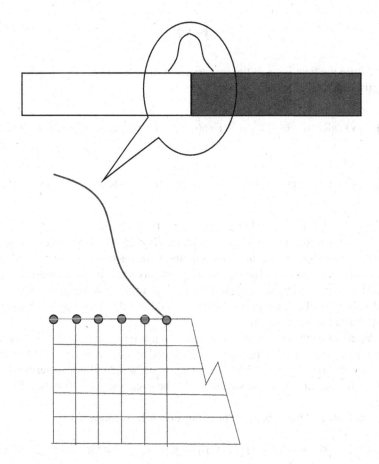

FIGURE 5.15 Discretization of the plate into various nodes and elements in the vicinity of the arc zone. The nodes which receive the arc heat are shown as bold dots.

where q is the heat flux in W/mm^2 which is experienced along the line and [N] is the shape function matrix. Since the arc heat input is experienced along a line, the shape function of a line element is used in the above equation.

The shape functions for a simplex line element are given as

$$N1 = (x2 - x) / (x2 - x1) \tag{5.10}$$

$$N2 = (x - x1) / (x2 - x1) \tag{5.11}$$

Using the above shape functions, the heat input terms are evaluated for the two nodes as

$$F1 = \int q\, N1.\, dx \tag{5.12}$$

$$F2 = \int q\, N2.\, dx \tag{5.13}$$

F1 and F2 are the heat input at the two nodes of the element.
Substituting the values of q, we get

$$F1 = \int 3\eta VI /\pi R^2.\, Exp(-3\, x^2/R^2).\, Exp(-3(v.\,(t-\tau))^2/R^2).\,(x2-x)/(x2-x1).\, dx \tag{5.14}$$

$$F2 = \int 3\eta VI /\pi R^2.\, Exp(-3\, x^2/R^2).\, Exp(-3(v.\,(t-\tau))^2/R^2).\,(x-x1)/(x2-x1).\, dx \tag{5.15}$$

The above expressions are evaluated for various time steps t.

The integration for the nodal heat input terms is performed by numerical integration method using Simpson's rule. In this method, the interval $x2 - x1$ is divided into n number of even segments. The number of segments into which the interval $x2 - x1$ is divided must be sufficiently high so that the final value shows convergence and the integral value calculated using n segments should not differ greatly from the value calculated using $n + 2$ segments.

The length of each segment is given as h which is equal to $(x2 - x1)/n$. There are $n + 1$ points in total, and the function is calculated at these $n + 1$ points. The values are denoted as F0, F1, F2 …. Fn. These values are weighted with a term which is equal to 1 for the first and last values, 4 for the even values and 2 for all the odd values.

The final integral value is given by Simpson's rule as follows:

$$F = h / 3.\, (F0 + 4.\, F1 + 2.\, F2 + \ldots + Fn) \tag{5.16}$$

The following worked out example shows the calculation procedure for a given case.

5.1.1 Example Problem: Calculate the arc heat values for various time steps for the following cross sectional case

Welding process = Submerged arc welding
Welding current I = 300 A
Welding voltage V = 32 V
Process efficiency = 1.0
Arc diameter = 10 mm
Welding speed = 5 mm/s
Plate thickness = 10 mm
Plate width = 300 mm
Distance moved by the arc for each time step = 1 mm

SOLUTION

Using a symmetrical model, a half section as shown in Figure 5.15 is taken for the analysis. The region under the arc is divided into 5 elements each having a length of 1 mm. The above calculation is performed for these five elements for any given time. The calculations must be repeated for different time steps from time t = 0 to time t = 2.τ. In the present case, the value of τ = arc radius/welding speed, i.e., 5/5 = 1 s.

The time increment is selected based on the distance moved by the arc for a time step. In the present case, the arc is assumed to move by 1 mm for each time step and since the welding speed is 5 mm/s, the corresponding time increment is 0.2 s. The section receives the arc heat for a duration of 2 s.

The arc heat is calculated for any time t as

$$F1 = \int 3\eta VI/\pi R^2 . \, Exp(-3\,x^2/R^2). \quad Exp\,(-3\,(v.\,(t-\tau))^2/R^2). \quad (x2-x)/(x2-x1).dx$$

$$F2 = \int 3\eta VI\,/\pi R^2 . \, Exp(-3\,x^2/R^2). \, Exp\,(-3\,(v.\,(t-\tau))^2/R^2).(x-x1)/(x2-x1).\,dx$$

To start with, the functions are evaluated for time t = 0. In the above equations, the value of t is taken as 0 and the calculations are made.

The total interval (x2 − x1) is divided into 10 segments, and x varies from x1 to x2 in steps of h, which is 0.1 mm in the present case. For each of the x values, the shape function and the heat flux are evaluated, and the product is multiplied by a weightage factor which is 1, 4, 2, 4, 2, 4, 2, 4, 2, 4 and 1 for the 11 values, respectively. The individual values are summed up and multiplied by h/3. The various terms in the element may be easily computed in a spreadsheet program like Excel. The tabular result of the computation is given in Table 5.1.

From the above, the values of F1 and F2 for the element 1 are calculated as 8.95 and 8.60 W, respectively.

The above procedure is repeated for the other elements 2, 3, 4 and 5. The results are presented in Tables 5.2 through 5.5.

TABLE 5.1
Calculation Table for Element 1

x1	x2	x	N1	N2	q	N1. q	N2. q	W	W.N1.q	W.N2.q
0.0	1.0	0	1.0	0	18.26	18.26	0	1	18.26	0
0.0	1.0	0.1	0.9	0.1	18.23	16.41	1.82	4	65.64	7.29
0.0	1.0	0.2	0.8	0.2	18.17	14.54	3.63	2	29.07	7.27
0.0	1.0	0.3	0.7	0.3	18.06	12.64	5.42	4	50.57	21.67
0.0	1.0	0.4	0.6	0.4	17.91	10.75	7.16	2	21.49	14.33
0.0	1.0	0.5	0.5	0.5	17.72	8.86	8.86	4	35.43	35.43
0.0	1.0	0.6	0.4	0.6	17.48	6.99	10.49	2	13.99	20.98
0.0	1.0	0.7	0.3	0.7	17.21	5.16	12.05	4	20.66	48.20
0.0	1.0	0.8	0.2	0.8	16.91	3.38	13.53	2	6.76	27.05
0.0	1.0	0.9	0.1	0.9	16.57	1.66	14.91	4	6.63	59.64
0.0	1.0	1.0	0	1.0	16.19	0	16.19	1	0	16.19
Total									268.50	258.06
Total x (h/3)									8.95	8.60

TABLE 5.2
Calculation Table for Element 2

x1	x2	x	N1	N2	q	N1. q	N2.q	W	W.N1.q	W.N2.q
1.0	2.0	1.0	1.0	0	16.19	16.19	0	1	16.19	0
1.0	2.0	1.1	0.9	0.1	15.79	14.21	1.58	4	56.84	6.32
1.0	2.0	1.2	0.8	0.2	15.36	12.29	3.07	2	24.57	6.14
1.0	2.0	1.3	0.7	0.3	14.91	10.43	4.47	4	41.74	17.89
1.0	2.0	1.4	0.6	0.4	14.43	8.66	5.77	2	17.32	11.54
1.0	2.0	1.5	0.5	0.5	13.94	6.97	6.97	4	27.87	27.87
1.0	2.0	1.6	0.4	0.6	13.43	5.37	8.06	2	10.74	16.11
1.0	2.0	1.7	0.3	0.7	12.91	3.87	9.03	4	15.49	36.14
1.0	2.0	1.8	0.2	0.8	12.38	2.48	9.90	2	4.95	19.80
1.0	2.0	1.9	0.1	0.9	11.84	1.18	10.65	4	4.74	42.62
1.0	2.0	2.0	0	1.0	11.30	0	11.30	1	0	11.30
Total									220.45	195.73
Total x (h/3)									7.35	6.52

TABLE 5.3
Calculation Table for Element 3

x1	x2	x	N1	N2	q	N1. q	N2.q	W	W.N1.q	W.N2.q
2.0	3.0	2.0	1.0	0	11.30	11.30	0	1	11.30	0
2.0	3.0	2.1	0.9	0.1	10.75	9.68	1.08	4	38.72	4.30
2.0	3.0	2.2	0.8	0.2	10.21	8.17	2.04	2	16.34	4.09
2.0	3.0	2.3	0.7	0.3	9.68	6.77	2.90	4	27.09	11.61

(Continued)

TABLE 5.3 (*Continued*)
Calculation Table for Element 3

x1	x2	x	N1	N2	q	N1. q	N2.q	W	W.N1.q	W.N2.q
2.0	3.0	2.4	0.6	0.4	9.15	5.49	3.66	2	10.98	7.32
2.0	3.0	2.5	0.5	0.5	8.62	4.31	4.31	4	17.25	17.25
2.0	3.0	2.6	0.4	0.6	8.11	3.24	4.87	2	6.49	9.73
2.0	3.0	2.7	0.3	0.7	7.61	2.28	5.33	4	9.13	21.31
2.0	3.0	2.8	0.2	0.8	7.13	1.43	5.70	2	2.85	11.40
2.0	3.0	2.9	0.1	0.9	6.65	0.67	5.99	4	2.66	23.96
2.0	3.0	3.0	0	1.0	6.20	0	6.20	1	0	6.20
Total									142.81	117.17
Total x (h/3)									4.76	3.91

TABLE 5.4
Calculation Table for Element 4

x1	x2	x	N1	N2	q	N1. q	N2.q	W	W.N1.q	W.N2.q
3.0	4.0	3.0	1.0	0	6.20	6.20	0	1	6.20	0
3.0	4.0	3.1	0.9	0.1	5.76	5.19	0.58	4	20.74	2.30
3.0	4.0	3.2	0.8	0.2	5.34	4.27	1.07	2	8.55	2.14
3.0	4.0	3.3	0.7	0.3	4.94	3.46	1.48	4	13.84	5.93
3.0	4.0	3.4	0.6	0.4	4.56	2.74	1.82	2	5.47	3.65
3.0	4.0	3.5	0.5	0.5	4.20	2.10	2.10	4	8.40	8.40
3.0	4.0	3.6	0.4	0.6	3.85	1.54	2.31	2	3.08	4.63
3.0	4.0	3.7	0.3	0.7	3.53	1.06	2.47	4	4.24	9.89
3.0	4.0	3.8	0.2	0.8	3.23	0.65	2.58	2	1.29	5.16
3.0	4.0	3.9	0.1	0.9	2.94	0.29	2.65	4	1.18	10.59
3.0	4.0	4.0	0	1.0	2.68	0	2.68	1	0	2.68
Total									72.99	55.36
Total x (h/3)									2.43	1.85

TABLE 5.5
Calculation Table for Element 5

x1	x2	x	N1	N2	q	N1. Q	N2.q	W	W.N1.q	W.N2.q
4.0	5.0	4.0	1.0	0	2.68	2.68	0	1	2.68	0
4.0	5.0	4.1	0.9	0.1	2.43	2.19	0.24	4	8.74	0.97
4.0	5.0	4.2	0.8	0.2	2.20	1.76	0.44	2	3.52	0.88
4.0	5.0	4.3	0.7	0.3	1.99	1.39	0.60	4	5.56	2.38
4.0	5.0	4.4	0.6	0.4	1.79	1.07	0.72	2	2.15	1.43
4.0	5.0	4.5	0.5	0.5	1.61	0.80	0.80	4	3.21	3.21
4.0	5.0	4.6	0.4	0.6	1.44	0.58	0.86	2	1.15	1.73

(*Continued*)

TABLE 5.5 (*Continued*)

Calculation Table for Element 5

x1	x2	x	N1	N2	q	N1. Q	N2.q	W	W.N1.q	W.N2.q
4.0	5.0	4.7	0.3	0.7	1.29	0.39	0.90	4	1.55	3.61
4.0	5.0	4.8	0.2	0.8	1.15	0.23	0.92	2	0.46	1.84
4.0	5.0	4.9	0.1	0.9	1.02	0.10	0.92	4	0.41	3.68
4.0	5.0	5.0	0	1.0	0.91	0	0.91	1	0	0.91
Total									29.43	20.65
Total x (h/3)									0.98	0.69

From the previous tables, the heat values for the five elements have been determined. For each element, two values have been computed which correspond to the heat at the nodes of the element as shown in Figure 5.16. The nodal heat values are totalled as shown in Table 5.6.

Thus, the heat liberated at the 6 nodes corresponding to a time t = 0 are 8.95, 15.95, 11.28, 6.34, 2.83, 0.69 W, respectively. The analysis can be repeated for different time intervals, and the heat input values for the six nodes can be determined. The calculated heat values for the various time steps are tabulated in Table 5.7.

From the above table, it can be seen that the heat input increases gradually from time t = 0 to a time t = 1 s and decreases gradually from time t = 1 s to time t = 2 s. The maximum heat is experienced corresponding to a time t = 1 s when the centre of the arc is directly over the plane.

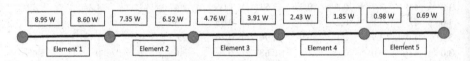

FIGURE 5.16 Heat distribution in the elements.

TABLE 5.6

Heat Input Values for Different Elements

Element	Node 1	Node 2	Node 3	Node 4	Node 5	Node 6
1	8.95	8.60				
2		7.35	6.52			
3			4.76	3.91		
4				2.43	1.85	
5					0.98	0.69
TOTAL	8.95	15.95	11.28	6.34	2.83	0.69

TABLE 5.7
Arc Heat Values for Various Time Steps

Time, s	Heat Input, watts					
	Node 1	Node 2	Node 3	Node 4	Node 5	Node 6
0	8.95	15.95	11.28	6.34	2.83	0.69
0.2	26.35	46.97	33.23	18.66	8.32	2.03
0.4	61.05	108.80	76.97	43.23	19.28	4.70
0.6	111.24	198.24	140.25	78.78	35.13	8.56
0.8	159.44	284.14	201.03	112.92	50.35	12.26
1.0	179.77	320.37	226.66	127.31	56.77	13.83
1.2	159.44	284.14	201.03	112.92	50.35	12.26
1.4	111.24	198.24	140.25	78.78	35.13	8.56
1.6	61.05	108.80	76.97	43.23	19.28	4.70
1.8	26.35	46.97	33.23	18.66	8.32	2.03
2.0	8.95	15.95	11.28	6.34	2.83	0.69

5.2 IN-PLANE ANALYSIS

The in-plane analysis is performed by considering a two-dimensional plane at the mid-section of the plate. The arc moves along the length of the plate from one end to another end as shown in Figure 5.17. At the start point, the arc is just touching one edge of the plate as shown in the figure. The arc moves to a newer location along the weld line as shown in Figure 5.18 for different time intervals. Finally the welding phase comes to an end when the arc just touches the other end of the plate

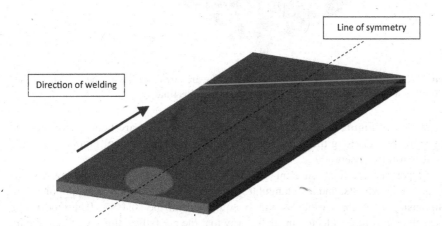

FIGURE 5.17 In-plane analysis of the plate. The arc circle is shown. The analysis starts from this time step (t = 0).

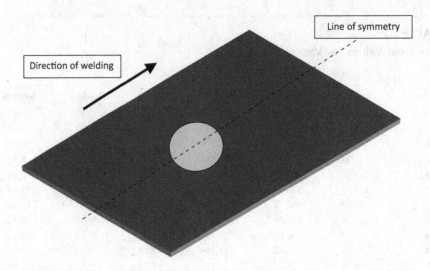

FIGURE 5.18   In-plane analysis of the plate. The arc circle is shown. The arc has travelled some distance along the plate.

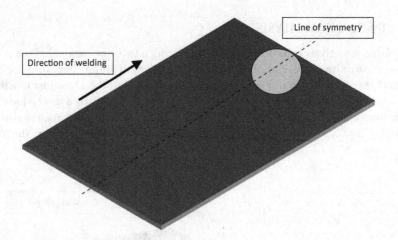

FIGURE 5.19   In-plane analysis of the plate. The arc circle is shown. The arc has reached the other end of the plate. The arc heat input ends at this time.

as shown in Figure 5.19. During the welding phase, the arc heat is experienced by some points along the weld line. During the subsequent cooling phase, the arc heat is totally absent.

To perform the finite element analysis, the component is discretized into various nodes and elements, and a rectangular four noded element is one of the popular two-dimensional elements which is widely employed for the analysis. If the mesh size near the arc zone is selected in such a way that the arc falls entirely within a rectangular element whose side is equal to the arc diameter as shown in Figure 5.20, then it is easy to surmise that the heat input at the four nodes will be equal to one fourth

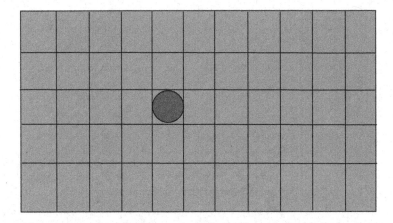

FIGURE 5.20 Finite element meshing of the plate with mesh size equal to the arc diameter.

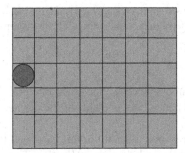

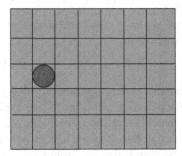

FIGURE 5.21 The location of the arc at successive time steps.

of the nVI value (product of process efficiency, the arc current and voltage). But if the element size is less than the arc diameter, then the calculations are more difficult.

A moving arc occupies different locations at different time intervals, and the simplest way of modelling the moving arc heat input is to assume that the arc moves to the next set of elements during the subsequent time step as shown in Figure 5.21.

This procedure is easy to implement as the same nodal heat values may be applied for the next set of nodes without the need to recalculate the arc heat for the new position. But this procedure puts a restriction on the time increment, which must be equal to the element size divided by the welding speed. If the element size is reasonably small, then this procedure will give good results. But if the element size is of the order of arc diameter then the time increment will be high, and this may lead to erratic and discontinuous thermal results. For getting a smooth result, the arc movement during a time step should be limited so that any point along the weld line will receive gradually increasing and gradually decreasing heat values. This will ensure that the temperature difference at any node for two successive time steps is not abnormally high and the thermal results are smooth.

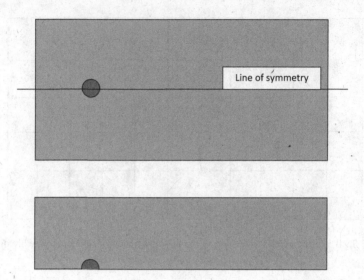

FIGURE 5.22   The top figure represents the actual size of the plate which is getting welded. The below picture shows a half model taking symmetry into consideration. The arc is represented as a circle.

In many instances where the arc is moving along the centreline of the component, the condition of symmetry can be applied and hence only one half of the arc needs to be modelled as shown in Figure 5.22.

The calculation of arc heat at various nodes is more difficult in this case, and the total heat applied at the various nodes should add up to half of the ηVI value. The arc movement must be gradual as shown in Figure 5.23 so that smooth thermal results may be obtained.

From the finite element analysis, the nodal values of heat input are obtained using the following relation

$$\{F\} = \int q\,[N]^{T}\,ds \tag{5.17}$$

where q is the heat flux which is experienced along the surface s and N represents the shape function matrix. The heat input is distributed in both x and y directions, hence the heat input at the four nodes has to be obtained by performing double integration with respect to x and y as follows.

$$F1 = \iint q.\,N1.\,dx.\,dy \tag{5.18}$$

$$F2 = \iint q.\,N2.\,dx.\,dy \tag{5.19}$$

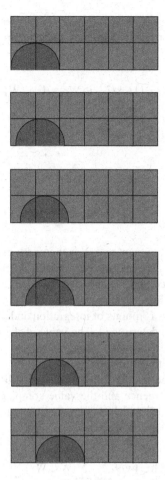

FIGURE 5.23 Movement of the arc along the weld line for different time steps. The gradual movement of the arc ensures a smooth thermal cycle.

$$F3 = \int\int q. \, N3. \, dx. \, dy \qquad (5.20)$$

$$F4 = \int\int q. \, N4. \, dx. \, dy \qquad (5.21)$$

The arc is assumed to move in the x direction, and the heat flux at any location within the element can be calculated using the relation

$$q \, (x,y,t) = 3\eta VI /\pi R^2. \, Exp \, (-3((x-v \, (t-\tau))^2/R^2). \, Exp \, (-3y^2/R^2) \qquad (5.22)$$

In the above relation, the time t is used to locate the arc in the element.

The shape functions for the rectangular element are given by the following relations.

$$N1 = \left(1 - x/L - y/W + x/L.\, y/W\right)$$

$$N2 = \left(x/L - x/L.\, y/W\right)$$

$$N3 = \left(x/L.\, y/W\right)$$

$$N4 = \left(y/W - x/L.\, y/W\right)$$

where L is the length of the element and W is the width of the element.

The double integration is performed by numerical integration method using Simpson's rule. The distance along x and y are divided into n number of divisions, and hx and hy are the values of increments in x and y directions, respectively. There will be a total of $(n + 1)^2$ points of integration and at each point, the heat flux and the shape functions are determined. The value of the function is weighted with a term as 1, 4, 2, 4, 2…., 2, 1 for x and y separately. The values are summed up and then multiplied by hx. hy/9 to get the final result.

As mentioned earlier, the number of divisions must be sufficiently high so that the final value shows convergence and the value computed using n segments does not differ much from the value which is computed using n + 2 segments.

The final integrated value is obtained as

$$F = hx.\, hy/9. \sum \sum Wx.\, Wy.\, N.\, q \qquad (5.23)$$

where hx is the increment in the x direction given as L/n where L is the length of the element in the x direction and n is the number of divisions;

hy is the increment in the y direction given as W/n where W is the width of the element in the y direction and n is the number of divisions;

Wx and Wy are the weightage factors given as 1, 4, 2, 4, 2…., 2, 1;

N is the shape function; and

q is the heat flux.

The calculations are performed for different time steps. At time t = 0, the arc is about to enter in the element. The time taken by the arc to cross an element is given as (element length + arc diameter)/welding speed. As mentioned earlier, the time increment employed in the analysis should correspond to the movement of the arc by a small value compared to the arc diameter. The total time taken by the arc to cross an element must be divided into various time steps, and for each value of time step, the calculations must be repeated.

5.2.1 Example Problem 1

Find the nodal heat input at the four nodes of a quadrilateral element for the following case.

Welding process = Gas metal arc welding
Current = 150 A
Voltage = 23 V
Process efficiency = 0.7
Arc diameter = 10 mm
Speed of welding = 5 mm/s
Length of the element = 10 mm
Width of the element = 5 mm
Time = 2 s

The rectangular element is of dimensions 10 × 5 mm as shown. At time t = 0, the arc is about to enter into the element, and at 2 s, the arc occupies the element as shown in Figure 5.24. At 4 s, the arc comes out of the element.

For calculating the arc heat at the four nodes of the rectangular element, the following integrals have to be evaluated using Simpson's rule.

$$F1 = \iint N1.3\eta VI/\pi R^2. \text{Exp}(-3(x - v(t - \tau))^2/R^2).\exp(-3y^2/R^2)\,dx.dy$$

$$F2 = \iint N2.3\eta VI/\pi R^2.\text{Exp}(-3(x - v(t - \tau))^2/R^2).\exp(-3y^2/R^2)dx.\,dy$$

$$F3 = \iint N3.3\eta VI/\pi R^2.\text{Exp}(-3(x - v(t - \tau))^2/R^2).\exp(-3y^2/R^2)dx.\,dy$$

$$F4 = \iint N4.3\eta VI/\pi R^2.\text{Exp}(-3(x - v(t - \tau))^2/R^2).\exp(-3y^2/R^2)\,dx.\,dy$$

The time t in the above relations is taken as 2s and the calculations are performed.

The length and width in the x and y directions are divided into 10 segments each, the values of increments hx and hy are given as

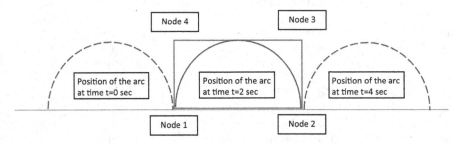

FIGURE 5.24 The position of the arc in the rectangular element for various time steps.

$$hx = (x2 - x1)/10$$

$$hy = (y2 - y1)/10$$

The weightage factors for each x and y values will be 1, 4, 2, 4, 2, 4, 2, 4, 2, 4, 1, respectively.

The values of x and y are incremented from the initial values of 0, 0 by hx and hy, respectively, and the function is evaluated at a total of 121 points. The final value of the integral is given as

$$F = hx. \ hy/9^* \sum \sum Wx. \ Wy.N.q$$

The calculations can be easily performed in a spreadsheet program like Excel. The results are shown in Tables 5.8 through 5.11 for the four nodes, respectively.

Thus the heat input at the four nodes in watts corresponding to time t = 2 s is as given in Figure 5.25.

It can be seen that the heat input is equal at node 1 and node 2. Similarly, the values are equal for node 3 and node 4. The total heat input for the four

---

TABLE 5.8
Calculation of Heat for Node 1 for t = 2 s

S. No.	x	y	N	q	Wx	Wy	hx	hy	N.q. Wx.Wy. hx.hy/9
1	0	0	1	4.59	1	1	1	0.5	0.26
2	0	0.5	0.9	4.46	1	4	1	0.5	0.89
3	0	1	0.8	4.07	1	2	1	0.5	0.36
4	0	1.5	0.7	3.51	1	4	1	0.5	0.55
5	0	2	0.6	2.84	1	2	1	0.5	0.19
6	0	2.5	0.5	2.17	1	4	1	0.5	0.24
7	0	3	0.4	1.56	1	2	1	0.5	0.07
8	0	3.5	0.3	1.06	1	4	1	0.5	0.07
9	0	4	0.2	0.67	1	2	1	0.5	0.01
10	0	4.5	0.1	0.40	1	4	1	0.5	0.01
11	0	5	0	0.23	1	1	1	0.5	0.0
12	1	0	0.9	13.52	4	1	1	0.5	2.70
13	1	0.5	0.81	13.12	4	4	1	0.5	9.45
14	1	1	0.72	11.99	4	2	1	0.5	3.84
15	1	1.5	0.63	10.32	4	4	1	0.5	5.78
16	1	2	0.54	8.37	4	2	1	0.5	2.01
17	1	2.5	0.45	6.39	4	4	1	0.5	2.56
18	1	3	0.36	4.59	4	2	1	0.5	0.73
19	1	3.5	0.27	3.11	4	4	1	0.5	0.75
20	1	4	0.18	1.98	4	2	1	0.5	0.16
21	1	4.5	0.09	1.19	4	4	1	0.5	0.10
22	1	5	0	0.67	4	1	1	0.5	0.0

(Continued)

TABLE 5.8 (*Continued*)
Calculation of Heat for Node 1 for t = 2 s

S. No.	x	y	N	q	Wx	Wy	hx	hy	N.q. Wx.Wy. hx.hy/9
23	2	0	0.8	31.33	2	1	1	0.5	2.78
24	2	0.5	0.72	30.40	2	4	1	0.5	9.73
25	2	1	0.64	27.78	2	2	1	0.5	3.95
26	2	1.5	0.56	23.91	2	4	1	0.5	5.95
27	2	2	0.48	19.38	2	2	1	0.5	2.07
28	2	2.5	0.4	14.80	2	4	1	0.5	2.63
29	2	3	0.32	10.64	2	2	1	0.5	0.76
30	2	3.5	0.24	7.20	2	4	1	0.5	0.77
31	2	4	0.16	4.59	2	2	1	0.5	0.16
32	2	4.5	0.08	2.76	2	4	1	0.5	0.10
33	2	5	0	1.56	2	1	1	0.5	0.0
34	3	0	0.7	57.08	4	1	1	0.5	8.88
35	3	0.5	0.63	55.39	4	4	1	0.5	31.02
36	3	1	0.56	50.63	4	2	1	0.5	12.60
37	3	1.5	0.49	43.57	4	4	1	0.5	18.98
38	3	2	0.42	35.32	4	2	1	0.5	6.59
39	3	2.5	0.35	26.96	4	4	1	0.5	8.39
40	3	3	0.28	19.38	4	2	1	0.5	2.41
41	3	3.5	0.21	13.12	4	4	1	0.5	2.45
42	3	4	0.14	8.37	4	2	1	0.5	0.52
43	3	4.5	0.07	5.03	4	4	1	0.5	0.31
44	3	5	0	2.84	4	1	1	0.5	0.0
45	4	0	0.6	81.82	2	1	1	0.5	5.45
46	4	0.5	0.54	79.40	2	4	1	0.5	19.06
47	4	1	0.48	72.56	2	2	1	0.5	7.74
48	4	1.5	0.42	62.46	2	4	1	0.5	11.66
49	4	2	0.36	50.63	2	2	1	0.5	4.05
50	4	2.5	0.3	38.65	2	4	1	0.5	5.15
51	4	3	0.24	27.78	2	2	1	0.5	1.48
52	4	3.5	0.18	18.81	2	4	1	0.5	1.50
53	4	4	0.12	11.99	2	2	1	0.5	0.32
54	4	4.5	0.06	7.20	2	4	1	0.5	0.19
55	4	5	0	4.07	2	1	1	0.5	0.0
56	5	0	0.5	92.25	4	1	1	0.5	10.25
57	5	0.5	0.45	89.52	4	4	1	0.5	35.81
58	5	1	0.4	81.82	4	2	1	0.5	14.54
59	5	1.5	0.35	70.42	4	4	1	0.5	21.91
60	5	2	0.3	57.08	4	2	1	0.5	7.61
61	5	2.5	0.25	43.57	4	4	1	0.5	9.68
62	5	3	0.2	31.33	4	2	1	0.5	2.78
63	5	3.5	0.15	21.21	4	4	1	0.5	2.83

(*Continued*)

TABLE 5.8 (*Continued*)
Calculation of Heat for Node 1 for t = 2 s

S. No.	x	y	N	q	Wx	Wy	hx	hy	N.q. Wx.Wy. hx.hy/9
64	5	4	0.1	13.52	4	2	1	0.5	0.60
65	5	4.5	0.05	8.12	4	4	1	0.5	0.36
66	5	5	0	4.59	4	1	1	0.5	0.0
67	6	0	0.4	81.82	2	1	1	0.5	3.64
68	6	0.5	0.36	79.40	2	4	1	0.5	12.70
69	6	1	0.32	72.56	2	2	1	0.5	5.16
70	6	1.5	0.28	62.46	2	4	1	0.5	7.77
71	6	2	0.24	50.63	2	2	1	0.5	2.70
72	6	2.5	0.2	38.65	2	4	1	0.5	3.44
73	6	3	0.16	27.78	2	2	1	0.5	0.99
74	6	3.5	0.12	18.81	2	4	1	0.5	1.0
75	6	4	0.08	11.99	2	2	1	0.5	0.21
76	6	4.5	0.04	7.20	2	4	1	0.5	0.13
77	6	5	0	4.07	2	1	1	0.5	0.0
78	7	0	0.3	57.08	4	1	1	0.5	3.81
79	7	0.5	0.27	55.39	4	4	1	0.5	13.29
80	7	1	0.24	50.63	4	2	1	0.5	5.40
81	7	1.5	0.21	43.57	4	4	1	0.5	8.13
82	7	2	0.18	35.32	4	2	1	0.5	2.83
83	7	2.5	0.15	26.96	4	4	1	0.5	3.60
84	7	3	0.12	19.38	4	2	1	0.5	1.03
85	7	3.5	0.09	13.12	4	4	1	0.5	1.05
86	7	4	0.06	8.37	4	2	1	0.5	0.22
87	7	4.5	0.03	5.03	4	4	1	0.5	0.13
88	7	5	0	2.84	4	1	1	0.5	0.0
89	8	0	0.2	31.33	2	1	1	0.5	0.70
90	8	0.5	0.18	30.40	2	4	1	0.5	2.43
91	8	1	0.16	27.78	2	2	1	0.5	0.99
92	8	1.5	0.14	23.91	2	4	1	0.5	1.49
93	8	2	0.12	19.38	2	2	1	0.5	0.52
94	8	2.5	0.1	14.80	2	4	1	0.5	0.66
95	8	3	0.08	10.64	2	2	1	0.5	0.19
96	8	3.5	0.06	7.20	2	4	1	0.5	0.19
97	8	4	0.04	4.59	2	2	1	0.5	0.04
98	8	4.5	0.02	2.76	2	4	1	0.5	0.02
99	8	5	0	1.56	2	1	1	0.5	0.0
100	9	0	0.1	13.52	4	1	1	0.5	0.30
101	9	0.5	0.09	13.12	4	4	1	0.5	1.05
102	9	1	0.08	11.99	4	2	1	0.5	0.43
103	9	1.5	0.07	10.32	4	4	1	0.5	0.64
104	9	2	0.06	8.37	4	2	1	0.5	0.22

(*Continued*)

TABLE 5.8 (*Continued*)
Calculation of Heat for Node 1 for t = 2 s

S. No.	x	y	N	q	Wx	Wy	hx	hy	N.q. Wx.Wy. hx.hy/9
105	9	2.5	0.05	6.39	4	4	1	0.5	0.28
106	9	3	0.04	4.59	4	2	1	0.5	0.08
107	9	3.5	0.03	3.11	4	4	1	0.5	0.08
108	9	4	0.02	1.98	4	2	1	0.5	0.02
109	9	4.5	0.01	1.19	4	4	1	0.5	0.01
110	9	5	0	0.67	4	1	1	0.5	0.0
111	10	0	0	4.59	1	1	1	0.5	0.0
112	10	0.5	0	4.46	1	4	1	0.5	0.0
113	10	1	0	4.07	1	2	1	0.5	0.0
114	10	1.5	0	3.51	1	4	1	0.5	0.0
115	10	2	0	2.84	1	2	1	0.5	0.0
116	10	2.5	0	2.17	1	4	1	0.5	0.0
117	10	3	0	1.56	1	2	1	0.5	0.0
118	10	3.5	0	1.06	1	4	1	0.5	0.0
119	10	4	0	0.67	1	2	1	0.5	0.0
120	10	4.5	0	0.40	1	4	1	0.5	0.0
121	10	5	0	0.23	1	1	1	0.5	0.0
Total									402.35

TABLE 5.9
Calculation of Heat for Node 2 for t = 2 s

S. No.	x	y	N	q	Wx	Wy	hx	hy	N.q. Wx.Wy. hx.hy/9
1	0	0	0	4.59	1	1	1	0.5	0.0
2	0	0.5	0	4.46	1	4	1	0.5	0.0
3	0	1	.0	4.07	1	2	1	0.5	0.0
4	0	1.5	0	3.51	1	4	1	0.5	0.0
5	0	2	0	2.84	1	2	1	0.5	0.0
6	0	2.5	0	2.17	1	4	1	0.5	0.0
7	0	3	0	1.56	1	2	1	0.5	0.0
8	0	3.5	0	1.06	1	4	1	0.5	0.0
9	0	4	0	0.67	1	2	1	0.5	0.0
10	0	4.5	0	0.40	1	4	1	0.5	0.0
11	0	5	0	0.23	1	1	1	0.5	0.0
12	1	0	0.1	13.52	4	1	1	0.5	0.30
13	1	0.5	0.09	13.12	4	4	1	0.5	1.05
14	1	1	0.08	11.99	4	2	1	0.5	0.43

(*Continued*)

TABLE 5.9 (*Continued*)
Calculation of Heat for Node 2 for t = 2 s

S. No.	x	y	N	q	Wx	Wy	hx	hy	N.q. Wx.Wy. hx.hy/9
15	1	1.5	0.07	10.32	4	4	1	0.5	0.64
16	1	2	0.06	8.37	4	2	1	0.5	0.22
17	1	2.5	0.05	6.39	4	4	1	0.5	0.28
18	1	3	0.04	4.59	4	2	1	0.5	0.08
19	1	3.5	0.03	3.11	4	4	1	0.5	0.08
20	1	4	0.02	1.98	4	2	1	0.5	0.02
21	1	4.5	0.01	1.19	4	4	1	0.5	0.01
22	1	5	0	0.67	4	1	1	0.5	0.0
23	2	0	0.2	31.33	2	1	1	0.5	0.70
24	2	0.5	0.18	30.40	2	4	1	0.5	2.43
25	2	1	0.16	27.78	2	2	1	0.5	0.99
26	2	1.5	0.14	23.91	2	4	1	0.5	1.49
27	2	2	0.12	19.38	2	2	1	0.5	0.52
28	2	2.5	0.1	14.80	2	4	1	0.5	0.66
29	2	3	0.08	10.64	2	2	1	0.5	0.19
30	2	3.5	0.06	7.20	2	4	1	0.5	0.19
31	2	4	0.04	4.59	2	2	1	0.5	0.04
32	2	4.5	0.02	2.76	2	4	1	0.5	0.02
33	2	5	0	1.56	2	1	1	0.5	0.0
34	3	0	0.3	57.08	4	1	1	0.5	3.81
35	3	0.5	0.27	55.39	4	4	1	0.5	13.29
36	3	1	0.24	50.63	4	2	1	0.5	5.40
37	3	1.5	0.21	43.57	4	4	1	0.5	8.13
38	3	2	0.18	35.32	4	2	1	0.5	2.83
39	3	2.5	0.15	26.96	4	4	1	0.5	3.60
40	3	3	0.12	19.38	4	2	1	0.5	1.03
41	3	3.5	0.09	13.12	4	4	1	0.5	1.05
42	3	4	0.06	8.37	4	2	1	0.5	0.22
43	3	4.5	0.03	5.03	4	4	1	0.5	0.13
44	3	5	0	2.84	4	1	1	0.5	0.0
45	4	0	0.4	81.82	2	1	1	0.5	3.64
46	4	0.5	0.36	79.40	2	4	1	0.5	12.70
47	4	1	0.32	72.56	2	2	1	0.5	5.16
48	4	1.5	0.28	62.46	2	4	1	0.5	7.77
49	4	2	0.24	50.63	2	2	1	0.5	2.70
50	4	2.5	0.2	38.65	2	4	1	0.5	3.44
51	4	3	0.16	27.78	2	2	1	0.5	0.99

(*Continued*)

TABLE 5.9 (*Continued*)

Calculation of Heat for Node 2 for t = 2 s

S. No.	x	y	N	q	Wx	Wy	hx	hy	N.q. Wx.Wy. hx.hy/9
52	4	3.5	0.12	18.81	2	4	1	0.5	1.0
53	4	4	0.08	11.99	2	2	1	0.5	0.21
54	4	4.5	0.04	7.20	2	4	1	0.5	0.13
55	4	5	0	4.07	2	1	1	0.5	0.0
56	5	0	0.5	92.25	4	1	1	0.5	10.25
57	5	0.5	0.45	89.52	4	4	1	0.5	35.81
58	5	1	0.4	81.82	4	2	1	0.5	14.54
59	5	1.5	0.35	70.42	4	4	1	0.5	21.91
60	5	2	0.3	57.08	4	2	1	0.5	7.61
61	5	2.5	0.25	43.57	4	4	1	0.5	9.68
62	5	3	0.2	31.33	4	2	1	0.5	2.78
63	5	3.5	0.15	21.21	4	4	1	0.5	2.83
64	5	4	0.1	13.52	4	2	1	0.5	0.60
65	5	4.5	0.05	8.12	4	4	1	0.5	0.36
66	5	5	0	4.59	4	1	1	0.5	0.0
67	6	0	0.6	81.82	2	1	1	0.5	5.45
68	6	0.5	0.54	79.40	2	4	1	0.5	19.06
69	6	1	0.48	72.56	2	2	1	0.5	7.74
70	6	1.5	0.42	62.46	2	4	1	0.5	11.66
71	6	2	0.36	50.63	2	2	1	0.5	4.05
72	6	2.5	0.3	38.65	2	4	1	0.5	5.15
73	6	3	0.24	27.78	2	2	1	0.5	1.48
74	6	3.5	0.18	18.81	2	4	1	0.5	1.50
75	6	4	0.12	11.99	2	2	1	0.5	0.32
76	6	4.5	0.06	7.20	2	4	1	0.5	0.19
77	6	5	0	4.07	2	1	1	0.5	0.0
78	7	0	0.7	57.08	4	1	1	0.5	8.88
79	7	0.5	0.63	55.39	4	4	1	0.5	31.02
80	7	1	0.56	50.63	4	2	1	0.5	12.60
81	7	1.5	0.49	43.57	4	4	1	0.5	18.98
82	7	2	0.42	35.32	4	2	1	0.5	6.59
83	7	2.5	0.35	26.96	4	4	1	0.5	8.39
84	7	3	0.28	19.38	4	2	1	0.5	2.41
85	7	3.5	0.21	13.12	4	4	1	0.5	2.45
86	7	4	0.14	8.37	4	2	1	0.5	0.52
87	7	4.5	0.07	5.03	4	4	1	0.5	0.31
88	7	5	0	2.84	4	1	1	0.5	0.0
89	8	0	0.8	31.33	2	1	1	0.5	2.78

(*Continued*)

TABLE 5.9 (*Continued*)
Calculation of Heat for Node 2 for t = 2 s

S. No.	x	y	N	q	Wx	Wy	hx	hy	N.q. Wx.Wy. hx.hy/9
90	8	0.5	0.72	30.40	2	4	1	0.5	9.73
91	8	1	0.64	27.78	2	2	1	0.5	3.95
92	8	1.5	0.56	23.91	2	4	1	0.5	5.95
93	8	2	0.48	19.38	2	2	1	0.5	2.07
94	8	2.5	0.4	14.80	2	4	1	0.5	2.63
95	8	3	0.32	10.64	2	2	1	0.5	0.76
96	8	3.5	0.24	7.20	2	4	1	0.5	0.77
97	8	4	0.16	4.59	2	2	1	0.5	0.16
98	8	4.5	0.08	2.76	2	4	1	0.5	0.10
99	8	5	0	1.56	2	1	1	0.5	0.0
100	9	0	0.9	13.52	4	1	1	0.5	2.70
101	9	0.5	0.81	13.12	4	4	1	0.5	9.45
102	9	1	0.72	11.99	4	2	1	0.5	3.84
103	9	1.5	0.63	10.32	4	4	1	0.5	5.78
104	9	2	0.54	8.37	4	2	1	0.5	2.01
105	9	2.5	0.45	6.39	4	4	1	0.5	2.56
106	9	3	0.36	4.59	4	2	1	0.5	0.73
107	9	3.5	0.27	3.11	4	4	1	0.5	0.75
108	9	4	0.18	1.98	4	2	1	0.5	0.16
109	9	4.5	0.09	1.19	4	4	1	0.5	0.10
110	9	5	0	0.67	4	1	1	0.5	0.0
111	10	0	1	4.59	1	1	1	0.5	0.26
112	10	0.5	0.9	4.46	1	4	1	0.5	0.89
113	10	1	0.8	4.07	1	2	1	0.5	0.36
114	10	1.5	0.7	3.51	1	4	1	0.5	0.55
115	10	2	0.6	2.84	1	2	1	0.5	0.19
116	10	2.5	0.5	2.17	1	4	1	0.5	0.24
117	10	3	0.4	1.56	1	2	1	0.5	0.07
118	10	3.5	0.3	1.06	1	4	1	0.5	0.07
119	10	4	0.2	0.67	1	2	1	0.5	0.01
120	10	4.5	0.1	0.40	1	4	1	0.5	0.01
121	10	5	0	0.23	1	1	1	0.5	0.0
Total									402.35

TABLE 5.10
Calculation of Heat for Node 3 for t = 2 s

S. No.	x	y	N	q	Wx	Wy	hx	hy	N.q. Wx.Wy. hx.hy/9
1	0	0	0	4.59	1	1	1	0.5	0.0
2	0	0.5	0	4.46	1	4	1	0.5	0.0
3	0	1	0	4.07	1	2	1	0.5	0.0
4	0	1.5	0	3.51	1	4	1	0.5	0.0
5	0	2	0	2.84	1	2	1	0.5	0.0
6	0	2.5	0	2.17	1	4	1	0.5	0.0
7	0	3	0	1.56	1	2	1	0.5	0.0
8	0	3.5	0	1.06	1	4	1	0.5	0.0
9	0	4	0	0.67	1	2	1	0.5	0.0
10	0	4.5	0	0.40	1	4	1	0.5	0.0
11	0	5	0	0.23	1	1	1	0.5	0.0
12	1	0	0	13.52	4	1	1	0.5	0.0
13	1	0.5	0.01	13.12	4	4	1	0.5	0.12
14	1	1	0.02	11.99	4	2	1	0.5	0.11
15	1	1.5	0.03	10.32	4	4	1	0.5	0.28
16	1	2	0.04	8.37	4	2	1	0.5	0.15
17	1	2.5	0.05	6.39	4	4	1	0.5	0.28
18	1	3	0.06	4.59	4	2	1	0.5	0.12
19	1	3.5	0.07	3.11	4	4	1	0.5	0.19
20	1	4	0.08	1.98	4	2	1	0.5	0.07
21	1	4.5	0.09	1.19	4	4	1	0.5	0.10
22	1	5	0.1	0.67	4	1	1	0.5	0.01
23	2	0	0	31.33	2	1	1	0.5	0.0
24	2	0.5	0.02	30.40	2	4	1	0.5	0.27
25	2	1	0.04	27.78	2	2	1	0.5	0.25
26	2	1.5	0.06	23.91	2	4	1	0.5	0.64
27	2	2	0.08	19.38	2	2	1	0.5	0.34
28	2	2.5	0.1	14.80	2	4	1	0.5	0.66
29	2	3	0.12	10.64	2	2	1	0.5	0.28
30	2	3.5	0.14	7.20	2	4	1	0.5	0.45
31	2	4	0.16	4.59	2	2	1	0.5	0.16
32	2	4.5	0.18	2.76	2	4	1	0.5	0.22
33	2	5	0.2	1.56	2	1	1	0.5	0.03
34	3	0	0	57.08	4	1	1	0.5	0.0
35	3	0.5	0.03	55.39	4	4	1	0.5	1.48
36	3	1	0.06	50.63	4	2	1	0.5	1.35
37	3	1.5	0.09	43.57	4	4	1	0.5	3.49
38	3	2	0.12	35.32	4	2	1	0.5	1.88
39	3	2.5	0.15	26.96	4	4	1	0.5	3.60
40	3	3	0.18	19.38	4	2	1	0.5	1.55
41	3	3.5	0.21	13.12	4	4	1	0.5	2.45

(Continued)

TABLE 5.10 (*Continued*)
Calculation of Heat for Node 3 for t = 2 s

S. No.	x	y	N	q	Wx	Wy	hx	hy	N.q. Wx.Wy. hx.hy/9
42	3	4	0.24	8.37	4	2	1	0.5	0.89
43	3	4.5	0.27	5.03	4	4	1	0.5	1.21
44	3	5	-0.3	2.84	4	1	1	0.5	0.19
45	4	0	0	81.82	2	1	1	0.5	0.0
46	4	0.5	0.04	79.40	2	4	1	0.5	1.41
47	4	1	0.08	72.56	2	2	1	0.5	1.29
48	4	1.5	0.12	62.46	2	4	1	0.5	3.33
49	4	2	0.16	50.63	2	2	1	0.5	1.80
50	4	2.5	0.2	38.65	2	4	1	0.5	3.44
51	4	3	0.24	27.78	2	2	1	0.5	1.48
52	4	3.5	0.28	18.81	2	4	1	0.5	2.34
53	4	4	0.32	11.99	2	2	1	0.5	0.85
54	4	4.5	0.36	7.20	2	4	1	0.5	1.15
55	4	5	0.4	4.07	2	1	1	0.5	0.18
56	5	0	0	92.25	4	1	1	0.5	0.0
57	5	0.5	0.05	89.52	4	4	1	0.5	3.98
58	5	1	0.1	81.82	4	2	1	0.5	3.64
59	5	1.5	0.15	70.42	4	4	1	0.5	9.39
60	5	2	0.2	57.08	4	2	1	0.5	5.07
61	5	2.5	0.25	43.57	4	4	1	0.5	9.68
62	5	3	0.3	31.33	4	2	1	0.5	4.18
63	5	3.5	0.35	21.21	4	4	1	0.5	6.60
64	5	4	0.4	13.52	4	2	1	0.5	2.40
65	5	4.5	0.45	8.12	4	4	1	0.5	3.25
66	5	5	0.5	4.59	4	1	1	0.5	0.51
67	6	0	0	81.82	2	1	1	0.5	0.0
68	6	0.5	0.06	79.40	2	4	1	0.5	2.12
69	6	1	0.12	72.56	2	2	1	0.5	1.94
70	6	1.5	0.18	62.46	2	4	1	0.5	5.0
71	6	2	0.24	50.63	2	2	1	0.5	2.70
72	6	2.5	0.3	38.65	2	4	1	0.5	5.15
73	6	3	0.36	27.78	2	2	1	0.5	2.22
74	6	3.5	0.42	18.81	2	4	1	0.5	3.51
75	6	4	0.48	11.99	2	2	1	0.5	1.28
76	6	4.5	0.54	7.20	2	4	1	0.5	1.73
77	6	5	0.6	4.07	2	1	1	0.5	0.27
78	7	0	0	57.08	4	1	1	0.5	0.0
79	7	0.5	0.07	55.39	4	4	1	0.5	3.45
80	7	1	0.14	50.63	4	2	1	0.5	3.15
81	7	1.5	0.21	43.57	4	4	1	0.5	8.13

(Continued)

TABLE 5.10 (*Continued*)
Calculation of Heat for Node 3 for t = 2 s

S. No.	x	y	N	q	Wx	Wy	hx	hy	N.q. Wx.Wy. hx.hy/9
82	7	2	0.28	35.32	4	2	1	0.5	4.40
83	7	2.5	0.35	26.96	4	4	1	0.5	8.39
84	7	3	0.42	19.38	4	2	1	0.5	3.62
85	7	3.5	0.49	13.12	4	4	1	0.5	5.72
86	7	4	0.56	8.37	4	2	1	0.5	2.08
87	7	4.5	0.63	5.03	4	4	1	0.5	2.81
88	7	5	0.7	2.84	4	1	1	0.5	0.44
89	8	0	0	31.33	2	1	1	0.5	0.0
90	8	0.5	0.08	30.40	2	4	1	0.5	1.08
91	8	1	0.16	27.78	2	2	1	0.5	0.99
92	8	1.5	0.24	23.91	2	4	1	0.5	2.55
93	8	2	0.32	19.38	2	2	1	0.5	1.38
94	8	2.5	0.4	14.80	2	4	1	0.5	2.63
95	8	3	0.48	10.64	2	2	1	0.5	1.13
96	8	3.5	0.56	7.20	2	4	1	0.5	1.79
97	8	4	0.64	4.59	2	2	1	0.5	0.65
98	8	4.5	0.72	2.76	2	4	1	0.5	0.88
99	8	5	0.8	1.56	2	1	1	0.5	0.14
100	9	0	0	13.52	4	1	1	0.5	0.0
101	9	0.5	0.09	13.12	4	4	1	0.5	1.05
102	9	1	0.18	11.99	4	2	1	0.5	0.96
103	9	1.5	0.27	10.32	4	4	1	0.5	2.48
104	9	2	0.36	8.37	4	2	1	0.5	1.34
105	9	2.5	0.45	6.39	4	4	1	0.5	2.56
106	9	3	0.54	4.59	4	2	1	0.5	1.10
107	9	3.5	0.63	3.11	4	4	1	0.5	1.74
108	9	4	0.72	1.98	4	2	1	0.5	0.63
109	9	4.5	0.81	1.19	4	4	1	0.5	0.86
110	9	5	0.9	0.67	4	1	1	0.5	0.13
111	10	0	0	4.59	1	1	1	0.5	0.0
112	10	0.5	0.1	4.46	1	4	1	0.5	0.10
113	10	1	0.2	4.07	i	2	1	0.5	0.09
114	10	1.5	0.3	3.51	1	4	1	0.5	0.23
115	10	2	0.4	2.84	1	2	1	0.5	0.13
116	10	2.5	0.5	2.17	1	4	1	0.5	0.24
117	10	3	0.6	1.56	1	2	1	0.5	0.10
118	10	3.5	0.7	1.06	1	4	1	0.5	0.16
119	10	4	0.8	0.67	1	2	1	0.5	0.06
120	10	4.5	0.9	0.40	1	4	1	0.5	0.08
121	10	5	1	0.23	1	1	1	0.5	0.01
Total									184.19

TABLE 5.11
Calculation of Heat for Node 4 for t = 2 s

S. No.	x	y	N	q	Wy	Wx	hx	hy	N.q. Wy.Wx. hx.hy/9
1	0	0	0	4.59	1	1	1	0.5	0.0
2	0	0.5	0.1	4.46	1	4	1	0.5	0.10
3	0	1	0.2	4.07	1	2	1	0.5	0.09
4	0	1.5	0.3	3.51	1	4	1	0.5	0.23
5	0	2	0.4	2.84	1	2	1	0.5	0.13
6	0	2.5	0.5	2.17	1	4	1	0.5	0.24
7	0	3	0.6	1.56	1	2	1	0.5	0.10
8	0	3.5	0.7	1.06	1	4	1	0.5	0.16
9	0	4	0.8	0.67	1	2	1	0.5	0.06
10	0	4.5	0.9	0.40	1	4	1	0.5	0.08
11	0	5	1	0.23	1	1	1	0.5	0.01
12	1	0	0	13.52	4	1	1	0.5	0.0
13	1	0.5	0.09	13.12	4	4	1	0.5	1.05
14	1	1	0.18	11.99	4	2	1	0.5	0.96
15	1	1.5	0.27	10.32	4	4	1	0.5	2.48
16	1	2	0.36	8.37	4	2	1	0.5	1.34
17	1	2.5	0.45	6.39	4	4	1	0.5	2.56
18	1	3	0.54	4.59	4	2	1	0.5	1.10
19	1	3.5	0.63	3.11	4	4	1	0.5	1.74
20	1	4	0.72	1.98	4	2	1	0.5	0.63
21	1	4.5	0.81	1.19	4	4	1	0.5	0.86
22	1	5	0.9	0.67	4	1	1	0.5	0.13
23	2	0	0	31.33	2	1	1	0.5	0.0
24	2	0.5	0.08	30.40	2	4	1	0.5	1.08
25	2	1	0.16	27.78	2	2	1	0.5	0.99
26	2	1.5	0.24	23.91	2	4	1	0.5	2.55
27	2	2	0.32	19.38	2	2	1	0.5	1.38
28	2	2.5	0.4	14.80	2	4	1	0.5	2.63
29	2	3	0.48	10.64	2	2	1	0.5	1.13
30	2	3.5	0.56	7.20	2	4	1	0.5	1.79
31	2	4	0.64	4.59	2	2	1	0.5	0.65
32	2	4.5	0.72	2.76	2	4	1	0.5	0.88
33	2	5	0.8	1.56	2	1	1	0.5	0.14
34	3	0	0	57.08	4	1	1	0.5	0.0
35	3	0.5	0.07	55.39	4	4	1	0.5	3.45
36	3	1	0.14	50.63	4	2	1	0.5	3.15
37	3	1.5	0.21	43.57	4	4	1	0.5	8.13
38	3	2	0.28	35.32	4	2	1	0.5	4.40
39	3	2.5	0.35	26.96	4	4	1	0.5	8.39
40	3	3	0.42	19.38	4	2	1	0.5	3.62

(Continued)

TABLE 5.11 (*Continued*)
Calculation of Heat for Node 4 for t = 2 s

S. No.	x	y	N	q	Wy	Wx	hx	hy	N.q. Wy.Wx. hx.hy/9
41	3	3.5	0.49	13.12	4	4	1	0.5	5.72
42	3	4	0.56	8.37	4	2	1	0.5	2.08
43	3	4.5	0.63	5.03	4	4	1	0.5	2.81
44	3	5	0.7	2.84	4	1	1	0.5	0.44
45	4	0	0	81.82	2	1	1	0.5	0.0
46	4	0.5	0.06	79.40	2	4	1	0.5	2.12
47	4	1	0.12	72.56	2	2	1	0.5	1.94
48	4	1.5	0.18	62.46	2	4	1	0.5	5.0
49	4	2	0.24	50.63	2	2	1	0.5	2.70
50	4	2.5	0.30	38.65	2	4	1	0.5	5.15
51	4	3	0.36	27.78	2	2	1	0.5	2.22
52	4	3.5	0.42	18.81	2	4	1	0.5	3.51
53	4	4	0.48	11.99	2	2	1	0.5	1.28
54	4	4.5	0.54	7.20	2	4	1	0.5	1.73
55	4	5	0.6	4.07	2	1	1	0.5	0.27
56	5	0	0	92.25	4	1	1	0.5	0.0
57	5	0.5	0.05	89.52	4	4	1	0.5	3.98
58	5	1	0.1	81.82	4	2	1	0.5	3.64
59	5	1.5	0.15	70.42	4	4	1	0.5	9.39
60	5	2	0.2	57.08	4	2	1	0.5	5.07
61	5	2.5	0.25	43.57	4	4	1	0.5	9.68
62	5	3	0.3	31.33	4	2	1	0.5	4.18
63	5	3.5	0.35	21.21	4	4	1	0.5	6.60
64	5	4	0.4	13.52	4	2	1	0.5	2.40
65	5	4.5	0.45	8.12	4	4	1	0.5	3.25
66	5	5	0.5	4.59	4	1	1	0.5	0.51
67	6	0	0	81.82	2	1	1	0.5	0.0
68	6	0.5	0.04	79.40	2	4	1	0.5	1.41
69	6	1	0.08	72.56	2	2	1	0.5	1.29
70	6	1.5	0.12	62.46	2	4	1	0.5	3.33
71	6	2	0.16	50.63	2	2	1	0.5	1.80
72	6	2.5	0.2	38.65	2	4	1	0.5	3.44
73	6	3	0.24	27.78	2	2	1	0.5	1.48
74	6	3.5	0.28	18.81	2	4	1	0.5	2.34
75	6	4	0.32	11.99	2	2	1	0.5	0.85
76	6	4.5	0.36	7.20	2	4	1	0.5	1.15
77	6	5	0.4	4.07	2	1	1	0.5	0.18
78	7	0	0	57.08	4	1	1	0.5	0.0
79	7	0.5	0.03	55.39	4	4	1	0.5	1.48
80	7	1	0.06	50.63	4	2	1	0.5	1.35
81	7	1.5	0.09	43.57	4	4	1	0.5	3.49

(*Continued*)

TABLE 5.11 (*Continued*)
Calculation of Heat for Node 4 for t = 2 s

S. No.	x	y	N	q	Wy	Wx	hx	hy	N.q. Wy.Wx. hx.hy/9
82	7	2	0.12	35.32	4	2	1	0.5	1.88
83	7	2.5	0.15	26.96	4	4	1	0.5	3.60
84	7	3	0.18	19.38	4	2	1	0.5	1.55
85	7	3.5	0.21	13.12	4	4	1	0.5	2.45
86	7	4	0.24	8.37	4	2	1	0.5	0.89
87	7	4.5	0.27	5.03	4	4	1	0.5	1.21
88	7	5	0.3	2.84	4	1	1	0.5	0.19
89	8	0	0	31.33	2	1	1	0.5	0.0
90	8	0.5	0.02	30.40	2	4	1	0.5	0.27
91	8	1	0.04	27.78	2	2	1	0.5	0.25
92	8	1.5	0.06	23.91	2	4	1	0.5	0.64
93	8	2	0.08	19.38	2	2	1	0.5	0.34
94	8	2.5	0.1	14.80	2	4	1	0.5	0.66
95	8	3	0.12	10.64	2	2	1	0.5	0.28
96	8	3.5	0.14	7.20	2	4	1	0.5	0.45
97	8	4	0.16	4.59	2	2	1	0.5	0.16
98	8	4.5	0.18	2.76	2	4	1	0.5	0.22
99	8	5	0.2	1.56	2	1	1	0.5	0.03
100	9	0	0	13.52	4	1	1	0.5	0.0
101	9	0.5	0.01	13.12	4	4	1	0.5	0.12
102	9	1	0.02	11.99	4	2	1	0.5	0.11
103	9	1.5	0.03	10.32	4	4	1	0.5	0.28
104	9	2	0.04	8.37	4	2	1	0.5	0.15
105	9	2.5	0.05	6.39	4	4	1	0.5	0.28
106	9	3	0.06	4.59	4	2	1	0.5	0.12
107	9	3.5	0.07	3.11	4	4	1	0.5	0.19
108	9	4	0.08	1.98	4	2	1	0.5	0.07
109	9	4.5	0.09	1.19	4	4	1	0.5	0.10
110	9	5	0.1	0.67	4	1	1	0.5	0.01
111	10	0	0	4.59	1	1	1	0.5	0.0
112	10	0.5	0	4.46	1	4	1	0.5	0.0
113	10	1	0	4.07	1	2	1	0.5	0.0
114	10	1.5	0	3.51	1	4	1	0.5	0.0
115	10	2	0	2.84	1	2	1	0.5	0.0
116	10	2.5	0	2.17	1	4	1	0.5	0.0
117	10	3	0	1.56	1	2	1	0.5	0.0
118	10	3.5	0	1.06	1	4	1	0.5	0.0
119	10	4	0	0.67	1	2	1	0.5	0.0
120	10	4.5	0	0.40	1	4	1	0.5	0.0
121	10	5	0	0.23	1	1	1	0.5	0.0
Total									184.19

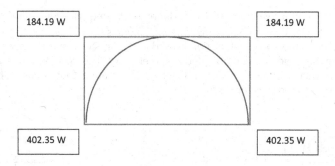

FIGURE 5.25 The arc heat values at the four nodes of the element at time t = 2 s.

nodes adds to 1173.07 W. The heat can also be calculated from the values of current, voltage, and process efficiency as 0.7 × 23 × 150 = 2415 W. Since half of the arc has been considered in the symmetric model, the total heat input should be 0.5 × 2415 = 1207.5 W. But the computation has yielded a total value of 1173.07 W instead of 1207.5 W as some stray value of arc heat falls outside the circular region and is not accounted in the calculation.

The calculations are repeated for various time intervals and the heat results for various time intervals are presented in Table 5.12. The time increment is selected

TABLE 5.12
Arc Heat Input for Various Time Steps

Time, s	Heat Input, watts			
	Node 1	Node 2	Node 3	Node 4
0	5.51	0.38	0.17	2.52
0.2	18.92	1.55	0.71	8.66
0.4	52.58	5.20	2.38	24.07
0.6	119.00	14.42	6.60	54.48
0.8	221.14	33.53	15.35	101.24
1.0	341.66	66.56	30.47	156.41
1.2	446.60	115.18	52.73	204.45
1.4	505.31	177.68	81.34	231.33
1.6	508.54	249.87	114.39	232.81
1.8	467.94	326.69	149.56	214.22
2.0	402.35	402.35	184.19	184.19
2.2	326.69	467.94	214.22	149.56
2.4	249.87	508.54	232.81	114.39
2.6	177.68	505.31	231.33	81.34
2.8	115.18	446.60	204.45	52.73
3.0	66.56	341.66	156.41	30.47
3.2	33.53	221.14	101.24	15.35
3.4	14.42	119.00	54.48	6.60
3.6	5.20	52.58	24.07	2.38
3.8	1.55	18.92	8.66	0.71
4.0	0.38	5.51	2.52	0.17

based on the arc movement by 1 mm during each time step. Since the welding speed is 5 mm/s, the corresponding time increment is 0.2 s.

The time in the above calculation is used for fixing the position of the arc and for determining the arc heat liberated in the element. The position of the arc for various time steps and the corresponding heat values at the four nodes of the element are presented in Figure 5.26.

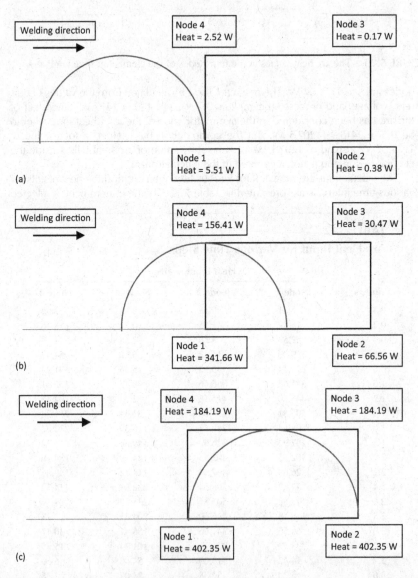

FIGURE 5.26 (a) The arc heat values at the four nodes of the element at time t = 0 s, (b) the arc heat values at the four nodes of the element at time t = 1 s, (c) the arc heat values at the four nodes of the element at time t = 2 s.   (*Continued*)

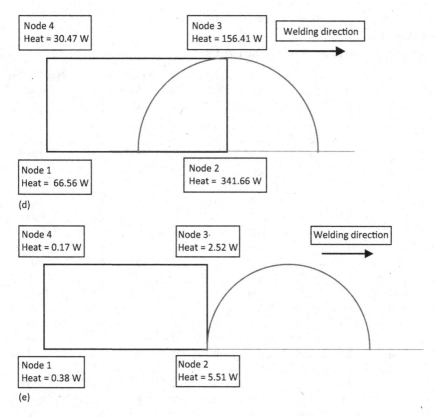

FIGURE 5.26 (Continued) (d) the arc heat values at the four nodes of the element at time t = 3 s, and (e) the arc heat values at the four nodes of the element at time t = 4 s.

5.2.2 Example Problem 2

Calculate the nodal heat values for the case of a gas metal arc welding using the following parameters.

Arc current = 100 A
Arc voltage = 20 V
Process efficiency = 0.75
Arc diameter = 10 mm
Welding speed = 5 mm/s
Length of the element = 5 mm
Width of the element = 5 mm
Distance moved by the arc for each time step = 1.0 mm

The calculation of arc heat for rectangular element of size 5 × 5 mm has been carried out as per the procedure explained earlier for time interval t = 1 s and the results for the four nodes are presented in Tables 5.13 through 5.16, respectively.

TABLE 5.13
Calculation of Heat for Node 1 for t = 1 s

S. No.	x	y	N	q	Wx	Wy	hx	hy	N.q. Wy.Wx. hx.hy/9
1	0	0	1	57.30	1	1	0.5	0.5	1.59
2	0	0.5	0.9	55.60	1	4	0.5	0.5	5.56
3	0	1	0.8	50.82	1	2	0.5	0.5	2.26
4	0	1.5	0.7	43.74	1	4	0.5	0.5	3.40
5	0	2	0.6	35.45	1	2	0.5	0.5	1.18
6	0	2.5	0.5	27.06	1	4	0.5	0.5	1.50
7	0	3	0.4	19.46	1	2	0.5	0.5	0.43
8	0	3.5	0.3	13.17	1	4	0.5	0.5	0.44
9	0	4	0.2	8.40	1	2	0.5	0.5	0.09
10	0	4.5	0.1	5.04	1	4	0.5	0.5	0.06
11	0	5	0	2.85	1	1	0.5	0.5	0.0
12	0.5	0	0.9	55.60	4	1	0.5	0.5	5.56
13	0.5	0.5	0.81	53.96	4	4	0.5	0.5	19.43
14	0.5	1	0.72	49.31	4	2	0.5	0.5	7.89
15	0.5	1.5	0.63	42.45	4	4	0.5	0.5	11.88
16	0.5	2	0.54	34.41	4	2	0.5	0.5	4.13
17	0.5	2.5	0.45	26.26	4	4	0.5	0.5	5.25
18	0.5	3	0.36	18.88	4	2	0.5	0.5	1.51
19	0.5	3.5	0.27	12.78	4	4	0.5	0.5	1.53
20	0.5	4	0.18	8.15	4	2	0.5	0.5	0.33
21	0.5	4.5	0.09	4.90	4	4	0.5	0.5	0.20
22	0.5	5	0	2.77	4	1	0.5	0.5	0.0
23	1	0	0.8	50.82	2	1	0.5	0.5	2.26
24	1	0.5	0.72	49.31	2	4	0.5	0.5	7.89
25	1	1	0.64	45.07	2	2	0.5	0.5	3.21
26	1	1.5	0.56	38.79	2	4	0.5	0.5	4.83
27	1	2	0.48	31.44	2	2	0.5	0.5	1.68
28	1	2.5	0.4	24.00	2	4	0.5	0.5	2.13
29	1	3	0.32	17.26	2	2	0.5	0.5	0.61
30	1	3.5	0.24	11.68	2	4	0.5	0.5	0.62
31	1	4	0.16	7.45	2	2	0.5	0.5	0.13
32	1	4.5	0.08	4.47	2	4	0.5	0.5	0.08
33	1	5	0	2.53	2	1	0.5	0.5	0.0
34	1.5	0	0.7	43.74	4	1	0.5	0.5	3.40
35	1.5	0.5	0.63	42.45	4	4	0.5	0.5	11.88
36	1.5	1	0.56	38.79	4	2	0.5	0.5	4.83
37	1.5	1.5	0.49	33.39	4	4	0.5	0.5	7.27
38	1.5	2	0.42	27.06	4	2	0.5	0.5	2.53
39	1.5	2.5	0.35	20.66	4	4	0.5	0.5	3.21
40	1.5	3	0.28	14.85	4	2	0.5	0.5	0.92
41	1.5	3.5	0.21	10.06	4	4	0.5	0.5	0.94

(Continued)

TABLE 5.13 (*Continued*)
Calculation of Heat for Node 1 for t = 1 s

S. No.	x	y	N	q	Wx	Wy	hx	hy	N.q. Wy.Wx. hx.hy/9
42	1.5	4	0.14	6.41	4	2	0.5	0.5	0.20
43	1.5	4.5	0.07	3.85	4	4	0.5	0.5	0.12
44	1.5	5	0	2.18	4	1	0.5	0.5	0.0
45	2	0	0.6	35.45	2	1	0.5	0.5	1.18
46	2	0.5	0.54	34.41	2	4	0.5	0.5	4.13
47	2	1	0.48	31.44	2	2	0.5	0.5	1.68
48	2	1.5	0.42	27.06	2	4	0.5	0.5	2.53
49	2	2	0.36	21.94	2	2	0.5	0.5	0.88
50	2	2.5	0.3	16.75	2	4	0.5	0.5	1.12
51	2	3	0.24	12.04	2	2	0.5	0.5	0.32
52	2	3.5	0.18	8.15	2	4	0.5	0.5	0.33
53	2	4	0.12	5.20	2	2	0.5	0.5	0.07
54	2	4.5	0.06	3.12	2	4	0.5	0.5	0.04
55	2	5	0	1.77	2	1	0.5	0.5	0.0
56	2.5	0	0.5	27.06	4	1	0.5	0.5	1.50
57	2.5	0.5	0.45	26.26	4	4	0.5	0.5	5.25
58	2.5	1	0.4	24.00	4	2	0.5	0.5	2.13
59	2.5	1.5	0.35	20.66	4	4	0.5	0.5	3.21
60	2.5	2	0.3	16.75	4	2	0.5	0.5	1.12
61	2.5	2.5	0.25	12.78	4	4	0.5	0.5	1.42
62	2.5	3	0.2	9.19	4	2	0.5	0.5	0.41
63	2.5	3.5	0.15	6.22	4	4	0.5	0.5	0.41
64	2.5	4	0.1	3.97	4	2	0.5	0.5	0.09
65	2.5	4.5	0.05	2.38	4	4	0.5	0.5	0.05
66	2.5	5	0	1.35	4	1	0.5	0.5	0.0
67	3	0	0.4	19.46	2	1	0.5	0.5	0.43
68	3	0.5	0.36	18.88	2	4	0.5	0.5.	1.51
69	3	1	0.32	17.26	2	2	0.5	0.5	0.61
70	3	1.5	0.28	14.85	2	4	0.5	0.5	0.92
71	3	2	0.24	12.04	2	2	0.5	0.5	0.32
72	3	2.5	0.2	9.19	2	4	0.5	0.5	0.41
73	3	3	0.16	6.61	2	2	0.5	0.5	0.12
74	3	3.5	0.12	4.47	2	4	0.5	0.5	0.12
75	3	4	0.08	2.85	2	2	0.5	0.5	0.03
76	3	4.5	0.04	1.71	2	4	0.5	0.5	0.02
77	3	5	0	0.97	2	1	0.5	0.5	0.0
78	3.5	0	0.3	13.17	4	1	0.5	0.5	0.44
79	3.5	0.5	0.27	12.78	4	4	0.5	0.5	1.53
80	3.5	1	0.24	11.68	4	2	0.5	0.5	0.62
81	3.5	1.5	0.21	10.06	4	4	0.5	0.5	0.94

(Continued)

TABLE 5.13 (*Continued*)
Calculation of Heat for Node 1 for t = 1 s

S. No.	x	y	N	q	Wx	Wy	hx	hy	N.q. Wy.Wx. hx.hy/9
82	3.5	2	0.18	8.15	4	2	0.5	0.5	0.33
83	3.5	2.5	0.15	6.22	4	4	0.5	0.5	0.41
84	3.5	3	0.12	4.47	4	2	0.5	0.5	0.12
85	3.5	3.5	0.09	3.03	4	4	0.5	0.5	0.12
86	3.5	4	0.06	1.93	4	2	0.5	0.5	0.03
87	3.5	4.5	0.03	1.16	4	4	0.5	0.5	0.02
88	3.5	5	0	0.66	4	1	0.5	0.5	0.0
89	4	0	0.2	8.40	2	1	0.5	0.5	0.09
90	4	0.5	0.18	8.15	2	4	0.5	0.5	0.33
91	4	1	0.16	7.45	2	2	0.5	0.5	0.13
92	4	1.5	0.14	6.41	2	4	0.5	0.5	0.20
93	4	2	0.12	5.20	2	2	0.5	0.5	0.07
94	4	2.5	0.1	3.97	2	4	0.5	0.5	0.09
95	4	3	0.08	2.85	2	2	0.5	0.5	0.03
96	4	3.5	0.06	1.93	2	4	0.5	0.5	0.03
97	4	4	0.04	1.23	2	2	0.5	0.5	0.01
98	4	4.5	0.02	0.74	2	4	0.5	0.5	0.0
99	4	5	0	0.42	2	1	0.5	0.5	0.0
100	4.5	0	0.1	5.04	4	1	0.5	0.5	0.06
101	4.5	0.5	0.09	4.90	4	4	0.5	0.5	0.20
102	4.5	1	0.08	4.47	4	2	0.5	0.5	0.08
103	4.5	1.5	0.07	3.85	4	4	0.5	0.5	0.12
104	4.5	2	0.06	3.12	4	2	0.5	0.5	0.04
105	4.5	2.5	0.05	2.38	4	4	0.5	0.5	0.05
106	4.5	3	0.04	1.71	4	2	0.5	0.5	0.02
107	4.5	3.5	0.03	1.16	4	4	0.5	0.5	0.02
108	4.5	4	0.02	0.74	4	2	0.5	0.5	0.0
109	4.5	4.5	0.01	0.44	4	4	0.5	0.5	0.0
110	4.5	5	0	0.25	4	1	0.5	0.5	0.0
111	5	0	0	2.85	1	1	0.5	0.5	0.0
112	5	0.5	0	2.77	1	4	0.5	0.5	0.0
113	5	1	0	2.53	1	2	0.5	0.5	0.0
114	5	1.5	0	2.18	1	4	0.5	0.5	0.0
115	5	2	0	1.77	1	2	0.5	0.5	0.0
116	5	2.5	0	1.35	1	4	0.5	0.5	0.0
117	5	3	0	0.97	1	2	0.5	0.5	0.0
118	5	3.5	0	0.66	1	4	0.5	0.5	0.0
119	5	4	0	0.42	1	2	0.5	0.5	0.0
120	5	4.5	0	0.25	1	4	0.5	0.5	0.0
121	5	5	0	0.14	1	1	0.5	0.5	0.0
Total									171.44

TABLE 5.14

Calculation of Heat for Node 2 for t = 1 s

S. No.	x	y	N	q	Wx	Wy	hx	hy	N.q. Wx.Wy. hx.hy/9
1	0	0	0	57.30	1	1	0.5	0.5	0.0
2	0	0.5	0	55.60	1	4	0.5	0.5	0.0
3	0	1	0	50.82	1	2	0.5	0.5	0.0
4	0	1.5	0	43.74	1	4	0.5	0.5	0.0
5	0	2	0	35.45	1	2	0.5	0.5	0.0
6	0	2.5	0	27.06	1	4	0.5	0.5	0.0
7	0	3	0	19.46	1	2	0.5	0.5	0.0
8	0	3.5	0	13.17	1	4	0.5	0.5	0.0
9	0	4	0	8.40	1	2	0.5	0.5	0.0
10	0	4.5	0	5.04	1	4	0.5	0.5	0.0
11	0	5	0	2.85	1	1	0.5	0.5	0.0
12	0.5	0	0.1	55.60	4	1	0.5	0.5	0.62
13	0.5	0.5	0.09	53.96	4	4	0.5	0.5	2.16
14	0.5	1	0.08	49.31	4	2	0.5	0.5	0.88
15	0.5	1.5	0.07	42.45	4	4	0.5	0.5	1.32
16	0.5	2	0.06	34.41	4	2	0.5	0.5	0.46
17	0.5	2.5	0.05	26.26	4	4	0.5	0.5	0.58
18	0.5	3	0.04	18.88	4	2	0.5	0.5	0.17
19	0.5	3.5	0.03	12.78	4	4	0.5	0.5	0.17
20	0.5	4	0.02	8.15	4	2	0.5	0.5	0.04
21	0.5	4.5	0.01	4.90	4	4	0.5	0.5	0.02
22	0.5	5	0	2.77	4	1	0.5	0.5	0.0
23	1	0	0.2	50.82	2	1	0.5	0.5	0.56
24	1	0.5	0.18	49.31	2	4	0.5	0.5	1.97
25	1	1	0.16	45.07	2	2	0.5	0.5	0.80
26	1	1.5	0.14	38.79	2	4	0.5	0.5	1.21
27	1	2	0.12	31.44	2	2	0.5	0.5	0.42
28	1	2.5	0.1	24.00	2	4	0.5	0.5	0.53
29	1	3	0.08	17.26	2	2	0.5	0.5	0.15
30	1	3.5	0.06	11.68	2	4	0.5	0.5	0.16
31	1	4	0.04	7.45	2	2	0.5	0.5	0.03
32	1	4.5	0.02	4.47	2	4	0.5	0.5	0.02
33	1	5	0	2.53	2	1	0.5	0.5	0.0
34	1.5	0	0.3	43.74	4	1	0.5	0.5	1.46
35	1.5	0.5	0.27	42.45	4	4	0.5	0.5	5.09
36	1.5	1	0.24	38.79	4	2	0.5	0.5	2.07
37	1.5	1.5	0.21	33.39	4	4	0.5	0.5	3.12
38	1.5	2	0.18	27.06	4	2	0.5	0.5	1.08
39	1.5	2.5	0.15	20.66	4	4	0.5	0.5	1.38
40	1.5	3	0.12	14.85	4	2	0.5	0.5	0.40
41	1.5	3.5	0.09	10.06	4	4	0.5	0.5	0.40

(Continued)

TABLE 5.14 (*Continued*)
Calculation of Heat for Node 2 for t = 1 s

S. No.	x	y	N	q	Wx	Wy	hx	hy	N.q. Wx.Wy. hx.hy/9
42	1.5	4	0.06	6.41	4	2	0.5	0.5	0.09
43	1.5	4.5	0.03	3.85	4	4	0.5	0.5	0.05
44	1.5	5	0	2.18	4	1	0.5	0.5	0.0
45	2	0	0.4	35.45	2	1	0.5	0.5	0.79
46	2	0.5	0.36	34.41	2	4	0.5	0.5	2.75
47	2	1	0.32	31.44	2	2	0.5	0.5	1.12
48	2	1.5	0.28	27.06	2	4	0.5	0.5	1.68
49	2	2	0.24	21.94	2	2	0.5	0.5	0.59
50	2	2.5	0.2	16.75	2	4	0.5	0.5	0.74
51	2	3	0.16	12.04	2	2	0.5	0.5	0.21
52	2	3.5	0.12	8.15	2	4	0.5	0.5	0.22
53	2	4	0.08	5.20	2	2	0.5	0.5	0.05
54	2	4.5	0.04	3.12	2	4	0.5	0.5	0.03
55	2	5	0	1.77	2	1	0.5	0.5	0.0
56	2.5	0	0.5	27.06	4	1	0.5	0.5	1.50
57	2.5	0.5	0.45	26.26	4	4	0.5	0.5	5.25
58	2.5	1	0.4	24.00	4	2	0.5	0.5	2.13
59	2.5	1.5	0.35	20.66	4	4	0.5	0.5	3.21
60	2.5	2	0.3	16.75	4	2	0.5	0.5	1.12
61	2.5	2.5	0.25	12.78	4	4	0.5	0.5	1.42
62	2.5	3	0.2	9.19	4	2	0.5	0.5	0.41
63	2.5	3.5	0.15	6.22	4	4	0.5	0.5	0.41
64	2.5	4	0.1	3.97	4	2	0.5	0.5	0.09
65	2.5	4.5	0.05	2.38	4	4	0.5	0.5	0.05
66	2.5	5	0	1.35	4	1	0.5	0.5	0.0
67	3	0	0.6	19.46	2	1	0.5	0.5	0.65
68	3	0.5	0.54	18.88	2	4	0.5	0.5	2.27
69	3	1	0.48	17.26	2	2	0.5	0.5	0.92
70	3	1.5	0.42	14.85	2	4	0.5	0.5	1.39
71	3	2	0.36	12.04	2	2	0.5	0.5	0.48
72	3	2.5	0.3	9.19	2	4	0.5	0.5	0.61
73	3	3	0.24	6.61	2	2	0.5	0.5	0.18
74	3	3.5	0.18	4.47	2	4	0.5	0.5	0.18
75	3	4	0.12	2.85	2	2	0.5	0.5	0.04
76	3	4.5	0.06	1.71	2	4	0.5	0.5	0.02
77	3	5	0	0.97	2	1	0.5	0.5	0.0
78	3.5	0	0.7	13.17	4	1	0.5	0.5	1.02
79	3.5	0.5	0.63	12.78	4	4	0.5	0.5	3.58
80	3.5	1	0.56	11.68	4	2	0.5	0.5	1.45
81	3.5	1.5	0.49	10.06	4	4	0.5	0.5	2.19

(Continued)

TABLE 5.14 (*Continued*)
Calculation of Heat for Node 2 for t = 1 s

S. No.	x	y	N	q	Wx	Wy	hx	hy	N.q. Wx.Wy. hx.hy/9
82	3.5	2	0.42	8.15	4	2	0.5	0.5	0.76
83	3.5	2.5	0.35	6.22	4	4	0.5	0.5	0.97
84	3.5	3	0.28	4.47	4	2	0.5	0.5	0.28
85	3.5	3.5	0.21	3.03	4	4	0.5	0.5	0.28
86	3.5	4	0.14	1.93	4	2	0.5	0.5	0.06
87	3.5	4.5	0.07	1.16	4	4	0.5	0.5	0.04
88	3.5	5	0	0.66	4	1	0.5	0.5	0.0
89	4	0	0.8	8.40	2	1	0.5	0.5	0.37
90	4	0.5	0.72	8.15	2	4	0.5	0.5	1.30
91	4	1	0.64	7.45	2	2	0.5	0.5	0.53
92	4	1.5	0.56	6.41	2	4	0.5	0.5	0.80
93	4	2	0.48	5.20	2	2	0.5	0.5	0.28
94	4	2.5	0.4	3.97	2	4	0.5	0.5	0.35
95	4	3	0.32	2.85	2	2	0.5	0.5	0.10
96	4	3.5	0.24	1.93	2	4	0.5	0.5	0.10
97	4	4	0.16	1.23	2	2	0.5	0.5	0.02
98	4	4.5	0.08	0.74	2	4	0.5	0.5	0.01
99	4	5	0	0.42	2	1	0.5	0.5	0.0
100	4.5	0	0.9	5.04	4	1	0.5	0.5	0.50
101	4.5	0.5	0.81	4.90	4	4	0.5	0.5	1.76
102	4.5	1	0.72	4.47	4	2	0.5	0.5	0.72
103	4.5	1.5	0.63	3.85	4	4	0.5	0.5	1.08
104	4.5	2	0.54	3.12	4	2	0.5	0.5	0.37
105	4.5	2.5	0.45	2.38	4	4	0.5	0.5	0.48
106	4.5	3	0.36	1.71	4	2	0.5	0.5	0.14
107	4.5	3.5	0.27	1.16	4	4	0.5	0.5	0.14
108	4.5	4	0.18	0.74	4	2	0.5	0.5	0.03
109	4.5	4.5	0.09	0.44	4	4	0.5	0.5	0.02
110	4.5	5	0	0.25	4	1	0.5	0.5	0.0
111	5	0	1	2.85	1	1	0.5	0.5	0.08
112	5	0.5	0.9	2.77	1	4	0.5	0.5	0.28
113	5	1	0.8	2.53	1	2	0.5	0.5	0.11
114	5	1.5	0.7	2.18	1	4	0.5	0.5	0.17
115	5	2	0.6	1.77	1	2	0.5	0.5	0.06
116	5	2.5	0.5	1.35	1	4	0.5	0.5	0.07
117	5	3	0.4	0.97	1	2	0.5	0.5	0.02
118	5	3.5	0.3	0.66	1	4	0.5	0.5	0.02
119	5	4	0.2	0.42	1	2	0.5	0.5	0.0
120	5	4.5	0.1	0.25	1	4	0.5	0.5	0.0
121	5	5	0	0.14	1	1	0.5	0.5	0.0
Total									78.49

TABLE 5.15

Calculation of Heat for Node 3 for t = 1 s

S. No.	x	y	N	q	Wx	Wy	hx	hy	N.q. Wx.Wy. hx.hy/9
1	0	0	0	57.30	1	1	0.5	0.5	0.0
2	0	0.5	0	55.60	1	4	0.5	0.5	0.0
3	0	1	0	50.82	1	2	0.5	0.5	0.0
4	0	1.5	0	43.74	1	4	0.5	0.5	0.0
5	0	2	0	35.45	1	2	0.5	0.5	0.0
6	0	2.5	0	27.06	1	4	0.5	0.5	0.0
7	0	3	0	19.46	1	2	0.5	0.5	0.0
8	0	3.5	0	13.17	1	4	0.5	0.5	0.0
9	0	4	0	8.40	1	2	0.5	0.5	0.0
10	0	4.5	0	5.04	1	4	0.5	0.5	0.0
11	0	5	0	2.85	1	1	0.5	0.5	0.0
12	0.5	0	0	55.60	4	1	0.5	0.5	0.0
13	0.5	0.5	0.01	53.96	4	4	0.5	0.5	0.24
14	0.5	1	0.02	49.31	4	2	0.5	0.5	0.22
15	0.5	1.5	0.03	42.45	4	4	0.5	0.5	0.57
16	0.5	2	0.04	34.41	4	2	0.5	0.5	0.31
17	0.5	2.5	0.05	26.26	4	4	0.5	0.5	0.58
18	0.5	3	0.06	18.88	4	2	0.5	0.5	0.25
19	0.5	3.5	0.07	12.78	4	4	0.5	0.5	0.40
20	0.5	4	0.08	8.15	4	2	0.5	0.5	0.14
21	0.5	4.5	0.09	4.90	4	4	0.5	0.5	0.20
22	0.5	5	0.1	2.77	4	1	0.5	0.5	0.03
23	1	0	0	50.82	2	1	0.5	0.5	0.0
24	1	0.5	0.02	49.31	2	4	0.5	0.5	0.22
25	1	1	0.04	45.07	2	2	0.5	0.5	0.20
26	1	1.5	0.06	38.79	2	4	0.5	0.5	0.52
27	1	2	0.08	31.44	2	2	0.5	0.5	0.28
28	1	2.5	0.1	24.00	2	4	0.5	0.5	0.53
29	1	3	0.12	17.26	2	2	0.5	0.5	0.23
30	1	3.5	0.14	11.68	2	4	0.5	0.5	0.36
31	1	4	0.16	7.45	2	2	0.5	0.5	0.13
32	1	4.5	0.18	4.47	2	4	0.5	0.5	0.18
33	1	5	0.2	2.53	2	1	0.5	0.5	0.03
34	1.5	0	0	43.74	4	1	0.5	0.5	0.0
35	1.5	0.5	0.03	42.45	4	4	0.5	0.5	0.57
36	1.5	1	0.06	38.79	4	2	0.5	0.5	0.52
37	1.5	1.5	0.09	33.39	4	4	0.5	0.5	1.34
38	1.5	2	0.12	27.06	4	2	0.5	0.5	0.72
39	1.5	2.5	0.15	20.66	4	4	0.5	0.5	1.38
40	1.5	3	0.18	14.85	4	2	0.5	0.5	0.59

(Continued)

TABLE 5.15 (*Continued*)
Calculation of Heat for Node 3 for t = 1 s

S. No.	x	y	N	q	Wx	Wy	hx	hy	N.q. Wx.Wy. hx.hy/9
41	1.5	3.5	0.21	10.06	4	4	0.5	0.5	0.94
42	1.5	4	0.24	6.41	4	2	0.5	0.5	0.34
43	1.5	4.5	0.27	3.85	4	4	0.5	0.5	0.46
44	1.5	5	0.3	2.18	4	1	0.5	0.5	0.07
45	2	0	0	35.45	2	1	0.5	0.5	0.0
46	2	0.5	0.04	34.41	2	4	0.5	0.5	0.31
47	2	1	0.08	31.44	2	2	0.5	0.5	0.28
48	2	1.5	0.12	27.06	2	4	0.5	0.5	0.72
49	2	2	0.16	21.94	2	2	0.5	0.5	0.39
50	2	2.5	0.2	16.75	2	4	0.5	0.5	0.74
51	2	3	0.24	12.04	2	2	0.5	0.5	0.32
52	2	3.5	0.28	8.15	2	4	0.5	0.5	0.51
53	2	4	0.32	5.20	2	2	0.5	0.5	0.18
54	2	4.5	0.36	3.12	2	4	0.5	0.5	0.25
55	2	5	0.4	1.77	2	1	0.5	0.5	0.04
56	2.5	0	0	27.06	4	1	0.5	0.5	0.0
57	2.5	0.5	0.05	26.26	4	4	0.5	0.5	0.58
58	2.5	1	0.1	24.00	4	2	0.5	0.5	• 0.53
59	2.5	1.5	0.15	20.66	4	4	0.5	0.5	1.38
60	2.5	2	0.2	16.75	4	2	0.5	0.5	0.74
61	2.5	2.5	0.25	12.78	4	4	0.5	0.5	1.42
62	2.5	3	0.3	9.19	4	2	0.5	0.5	0.61
63	2.5	3.5	0.35	6.22	4	4	0.5	0.5	0.97
64	2.5	4	0.4	3.97	4	2	0.5	0.5	0.35
65	2.5	4.5	0.45	2.38	4	4	0.5	0.5	0.48
66	2.5	5	0.5	1.35	4	1	0.5	0.5	0.07
67	3	0	0	19.46	2	1	0.5	0.5	0.0
68	3	0.5	0.06	18.88	2	4	0.5	0.5	0.25
69	3	1	0.12	17.26	2	2	0.5	0.5	0.23
70	3	1.5	0.18	14.85	2	4	0.5	0.5	0.59
71	3	2	0.24	12.04	2	2	0.5	0.5	0.32
72	3	2.5	0.3	9.19	2	4	0.5	0.5	0.61
73	3	3	0.36	6.61	2	2	0.5	0.5	0.26
74	3	3.5	0.42	4.47	2	4	0.5	0.5	0.42
75	3	4	0.48	2.85	2	2	0.5	0.5	0.15
76	3	4.5	0.54	1.71	2	4	0.5	0.5	0.21
77	3	5	0.6	0.97	2	1	0.5	0.5	0.03
78	3.5	0	0	13.17	4	1	0.5	0.5	0.0
79	3.5	0.5	0.07	12.78	4	4	0.5	0.5	0.40
80	3.5	1	0.14	11.68	4	2	0.5	0.5	0.36

(Continued)

TABLE 5.15 (*Continued*)
Calculation of Heat for Node 3 for t = 1 s

S. No.	x	y	N	q	Wx	Wy	hx	hy	N.q. Wx.Wy. hx.hy/9
81	3.5	1.5	0.21	10.06	4	4	0.5	0.5	0.94
82	3.5	2	0.28	8.15	4	2	0.5	0.5	0.51
83	3.5	2.5	0.35	6.22	4	4	0.5	0.5	0.97
84	3.5	3	0.42	4.47	4	2	0.5	0.5	0.42
85	3.5	3.5	0.49	3.03	4	4	0.5	0.5	0.66
86	3.5	4	0.56	1.93	4	2	0.5	0.5	0.24
87	3.5	4.5	0.63	1.16	4	4	0.5	0.5	0.32
88	3.5	5	0.7	0.66	4	1	0.5	0.5	0.05
89	4	0	0	8.40	2	1	0.5	0.5	0.0
90	4	0.5	0.08	8.15	2	4	0.5	0.5	0.14
91	4	1	0.16	7.45	2	2	0.5	0.5	0.13
92	4	1.5	0.24	6.41	2	4	0.5	0.5	0.34
93	4	2	0.32	5.20	2	2	0.5	0.5	0.18
94	4	2.5	0.4	3.97	2	4	0.5	0.5	0.35
95	4	3	0.48	2.85	2	2	0.5	0.5	0.15
96	4	3.5	0.56	1.93	2	4	0.5	0.5	0.24
97	4	4	0.64	1.23	2	2	0.5	0.5	0.09
98	4	4.5	0.72	0.74	2	4	0.5	0.5	0.12
99	4	5	0.8	0.42	2	1	0.5	0.5	0.02
100	4.5	0	0	5.04	4	1	0.5	0.5	0.0
101	4.5	0.5	0.09	4.90	4	4	0.5	0.5	0.20
102	4.5	1	0.18	4.47	4	2	0.5	0.5	0.18
103	4.5	1.5	0.27	3.85	4	4	0.5	0.5	0.46
104	4.5	2	0.36	3.12	4	2	0.5	0.5	0.25
105	4.5	2.5	0.45	2.38	4	4	0.5	0.5	0.48
106	4.5	3	0.54	1.71	4	2	0.5	0.5	0.21
107	4.5	3.5	0.63	1.16	4	4	0.5	0.5	0.32
108	4.5	4	0.72	0.74	4	2	0.5	0.5	0.12
109	4.5	4.5	0.81	0.44	4	4	0.5	0.5	0.16
110	4.5	5	0.9	0.25	4	1	0.5	0.5	0.03
111	5	0	0	2.85	1	1	0.5	0.5	0.0
112	5	0.5	0.1	2.77	1	4	0.5	0.5	0.03
113	5	1	0.2	2.53	1	2	0.5	0.5	0.03
114	5	1.5	0.3	2.18	1	4	0.5	0.5	0.07
115	5	2	0.4	1.77	1	2	0.5	0.5	0.04
116	5	2.5	0.5	1.35	1	4	0.5	0.5	0.07
117	5	3	0.6	0.97	1	2	0.5	0.5	0.03
118	5	3.5	0.7	0.66	1	4	0.5	0.5	0.05
119	5	4	0.8	0.42	1	2	0.5	0.5	0.02
120	5	4.5	0.9	0.25	1	4	0.5	0.5	0.03
121	5	5	1	0.14	1	1	0.5	0.5	0.0
Total									35.93

TABLE 5.16
Calculation of Heat for Node 4 for t = 1 s

S. No.	x	y	N	q	Wx	Wy	hx	hy	N.q. Wx.Wy. hx.hy/9
1	0	0	0	57.30	1	1	0.5	0.5	0.0
2	0	0.5	0.1	55.60	1	4	0.5	0.5	0.62
3	0	1	0.2	50.82	1	2	0.5	0.5	0.56
4	0	1.5	0.3	43.74	1	4	0.5	0.5	1.46
5	0	2	0.4	35.45	1	2	0.5	0.5	0.79
6	0	2.5	0.5	27.06	1	4	0.5	0.5	1.50
7	0	3	0.6	19.46	1	2	0.5	0.5	0.65
8	0	3.5	0.7	13.17	1	4	0.5	0.5	1.02
9	0	4	0.8	8.40	1	2	0.5	0.5	0.37
10	0	4.5	0.9	5.04	1	4	0.5	0.5	0.50
11	0	5	1	2.85	1	1	0.5	0.5	0.08
12	0.5	0	0	55.60	4	1	0.5	0.5	0.0
13	0.5	0.5	0.09	53.96	4	4	0.5	0.5	2.16
14	0.5	1	0.18	49.31	4	2	0.5	0.5	1.97
15	0.5	1.5	0.27	42.45	4	4	0.5	0.5	5.09
16	0.5	2	0.36	34.41	4	2	0.5	0.5	2.75
17	0.5	2.5	0.45	26.26	4	4	0.5	0.5	5.25
18	0.5	3	0.54	18.88	4	2	0.5	0.5	2.27
19	0.5	3.5	0.63	12.78	4	4	0.5	0.5	3.58
20	0.5	4	0.72	8.15	4	2	0.5	0.5	1.30
21	0.5	4.5	0.81	4.90	4	4	0.5	0.5	1.76
22	0.5	5	0.9	2.77	4	1	0.5	0.5	0.28
23	1	0	0	50.82	2	1	0.5	0.5	0.0
24	1	0.5	0.08	49.31	2	4	0.5	0.5	0.88
25	1	1	0.16	45.07	2	2	0.5	0.5	0.80
26	1	1.5	0.24	38.79	2	4	0.5	0.5	2.07
27	1	2	0.32	31.44	2	2	0.5	0.5	1.12
28	1	2.5	0.4	24.00	2	4	0.5	0.5	2.13
29	1	3	0.48	17.26	2	2	0.5	0.5	0.92
30	1	3.5	0.56	11.68	2	4	0.5	0.5	1.45
31	1	4	0.64	7.45	2	2	0.5	0.5	0.53
32	1	4.5	0.72	4.47	2	4	0.5	0.5	0.72
33	1	5	0.8	2.53	2	1	0.5	0.5	0.11
34	1.5	0	0	43.74	4	1	0.5	0.5	0.0
35	1.5	0.5	0.07	42.45	4	4	0.5	0.5	1.32
36	1.5	1	0.14	38.79	4	2	0.5	0.5	1.21
37	1.5	1.5	0.21	33.39	4	4	0.5	0.5	3.12
38	1.5	2	0.28	27.06	4	2	0.5	0.5	1.68
39	1.5	2.5	0.35	20.66	4	4	0.5	0.5	3.21

(Continued)

TABLE 5.16 (*Continued*)
Calculation of Heat for Node 4 for t = 1 s

S. No.	x	y	N	q	Wx	Wy	hx	hy	N.q. Wx.Wy. hx.hy/9
40	1.5	3	0.42	14.85	4	2	0.5	0.5	1.39
41	1.5	3.5	0.49	10.06	4	4	0.5	0.5	2.19
42	1.5	4	0.56	6.41	4	2	0.5	0.5	0.80
43	1.5	4.5	0.63	3.85	4	4	0.5	0.5	1.08
44	1.5	5	0.7	2.18	4	1	0.5	0.5	0.17
45	2	0	0	35.45	2	1	0.5	0.5	0.0
46	2	0.5	0.06	34.41	2	4	0.5	0.5	0.46
47	2	1	0.12	31.44	2	2	0.5	0.5	0.42
48	2	1.5	0.18	27.06	2	4	0.5	0.5	1.08
49	2	2	0.24	21.94	2	2	0.5	0.5	0.59
50	2	2.5	0.30	16.75	2	4	0.5	0.5	1.12
51	2	3	0.36	12.04	2	2	0.5	0.5	0.48
52	2	3.5	0.42	8.15	2	4	0.5	0.5	0.76
53	2	4	0.48	5.20	2	2	0.5	0.5	0.28
54	2	4.5	0.54	3.12	2	4	0.5	0.5	0.37
55	2	5	0.6	1.77	2	1	0.5	0.5	0.06
56	2.5	0	0	27.06	4	1	0.5	0.5	0.0
57	2.5	0.5	0.05	26.26	4	4	0.5	0.5	0.58
58	2.5	1	0.1	24.00	4	2	0.5	0.5	0.53
59	2.5	1.5	0.15	20.66	4	4	0.5	0.5	1.38
60	2.5	2	0.2	16.75	4	2	0.5	0.5	0.74
61	2.5	2.5	0.25	12.78	4	4	0.5	0.5	1.42
62	2.5	3	0.3	9.19	4	2	0.5	0.5	0.61
63	2.5	3.5	0.35	6.22	4	4	0.5	0.5	0.97
64	2.5	4	0.4	3.97	4	2	0.5	0.5	0.35
65	2.5	4.5	0.45	2.38	4	4	0.5	0.5	0.48
66	2.5	5	0.5	1.35	4	1	0.5	0.5	0.07
67	3	0	0	19.46	2	1	0.5	0.5	0.0
68	3	0.5	0.04	18.88	2	4	0.5	0.5	0.17
69	3	1	0.08	17.26	2	2	0.5	0.5	0.15
70	3	1.5	0.12	14.85	2	4	0.5	0.5	0.40
71	3	2	0.16	12.04	2	2	0.5	0.5	0.21
72	3	2.5	0.2	9.19	2	4	0.5	0.5	0.41
73	3	3	0.24	6.61	2	2	0.5	0.5	0.18
74	3	3.5	0.28	4.47	2	4	0.5	0.5	0.28
75	3	4	0.32	2.85	2	2	0.5	0.5	0.10
76	3	4.5	0.36	1.71	2	4	0.5	0.5	0.14
77	3	5	0.4	0.97	2	1	0.5	0.5	0.02

(*Continued*)

TABLE 5.16 (*Continued*)
Calculation of Heat for Node 4 for t = 1 s

S. No.	x	y	N	q	Wx	Wy	hx	hy	N.q. Wx.Wy. hx.hy/9
78	3.5	0	0	13.17	4	1	0.5	0.5	0.0
79	3.5	0.5	0.03	12.78	4	4	0.5	0.5	0.17
80	3.5	1	0.06	11.68	4	2	0.5	0.5	0.16
81	3.5	1.5	0.09	10.06	4	4	0.5	0.5	0.40
82	3.5	2	0.12	8.15	4	2	0.5	0.5	0.22
83	3.5	2.5	0.15	6.22	4	4	0.5	0.5	0.41
84	3.5	3	0.18	4.47	4	2	0.5	0.5	0.18
85	3.5	3.5	0.21	3.03	4	4	0.5	0.5	0.28
86	3.5	4	0.24	1.93	4	2	0.5	0.5	0.10
87	3.5	4.5	0.27	1.16	4	4	0.5	0.5	0.14
88	3.5	5	0.3	0.66	4	1	0.5	0.5	0.02
89	4	0	0	8.40	2	1	0.5	0.5	0.0
90	4	0.5	0.02	8.15	2	4	0.5	0.5	0.04
91	4	1	0.04	7.45	2	2	0.5	0.5	0.03
92	4	1.5	0.06	6.41	2	4	0.5	0.5	0.09
93	4	2	0.08	5.20	2	2	0.5	0.5	0.05
94	4	2.5	0.1	3.97	2	4	0.5	0.5	0.09
95	4	3	0.12	2.85	2	2	0.5	0.5	0.04
96	4	3.5	0.14	1.93	2	4	0.5	0.5	0.06
97	4	4	0.16	1.23	2	2	0.5	0.5	0.02
98	4	4.5	0.18	0.74	2	4	0.5	0.5	0.03
99	4	5	0.2	0.42	2	1	0.5	0.5	0.0
100	4.5	0	0	5.04	4	1	0.5	0.5	0.0
101	4.5	0.5	0.01	4.90	4	4	0.5	0.5	0.02
102	4.5	1	0.02	4.47	4	2	0.5	0.5	0.02
103	4.5	1.5	0.03	3.85	4	4	0.5	0.5	0.05
104	4.5	2	0.04	3.12	4	2	0.5	0.5	0.03
105	4.5	2.5	0.05	2.38	4	4	0.5	0.5	0.05
106	4.5	3	0.06	1.71	4	2	0.5	0.5	0.02
107	4.5	3.5	0.07	1.16	4	4	0.5	0.5	0.04
108	4.5	4	0.08	0.74	4	2	0.5	0.5	0.01
109	4.5	4.5	0.09	0.44	4	4	0.5	0.5	0.02
110	4.5	5	0.1	0.25	4	1	0.5	0.5	0.0
111	5	0	0	2.85	1	1	0.5	0.5	0.0
112	5	0.5	0	2.77	1	4	0.5	0.5	0.0
113	5	1	0	2.53	1	2	0.5	0.5	0.0
114	5	1.5	0	2.18	1	4	0.5	0.5	0.0
115	5	2	0	1.77	1	2	0.5	0.5	0.0
116	5	2.5	0	1.35	1	4	0.5	0.5	0.0

(*Continued*)

TABLE 5.16 (*Continued*)
Calculation of Heat for Node 4 for t = 1 s

S. No.	x	y	N	q	Wx	Wy	hx	hy	N.q. Wx.Wy. hx.hy/9
117	5	3	0	0.97	1	2	0.5	0.5	0.0
118	5	3.5	0	0.66	1	4	0.5	0.5	0.0
119	5	4	0	0.42	1	2	0.5	0.5	0.0
120	5	4.5	0	0.25	1	4	0.5	0.5	0.0
121	5	5	0	0.14	1	1	0.5	0.5	0.0
Total									78.49

The calculations are repeated for various time intervals and the heat results for various time intervals are presented in Table 5.17.

The heat liberated at the four nodes of rectangular element is presented in Figure 5.27.

TABLE 5.17
Arc Heat Values for Various Time Steps

	Heat Input, watts			
Time, s	Node 1	Node 2	Node 3	Node 4
0	3.15	0.48	0.22	1.44
0.2	10.73	1.95	0.89	4.91
0.4	29.42	6.47	2.96	13.47
0.6	65.05	17.75	8.13	29.78
0.8	116.76	40.67	18.62	53.45
1.0	171.44	78.49	35.93	78.49
1.2	207.78	128.38	58.77	95.12
1.4	209.89	178.36	81.65	96.08
1.6	178.36	209.89	96.08	81.65
1.8	128.38	207.78	95.12	58.77
2.0	78.49	171.44	78.49	35.93
2.2	40.67	116.76	53.45	18.62
2.4	17.75	65.05	29.78	8.13
2.6	6.47	29.42	13.47	2.96
2.8	1.95	10.73	4.91	0.89
3.0	0.48	3.15	1.44	0.22

5.2.3 Example Problem 3

Calculate the nodal heat values for the following case.

Welding process = Gas tungsten arc welding
Current = 80 A
Voltage = 12 V
Process efficiency = 0.6
Arc diameter = 4 mm
Welding speed = 1 mm/s
Length of the element = 5 mm

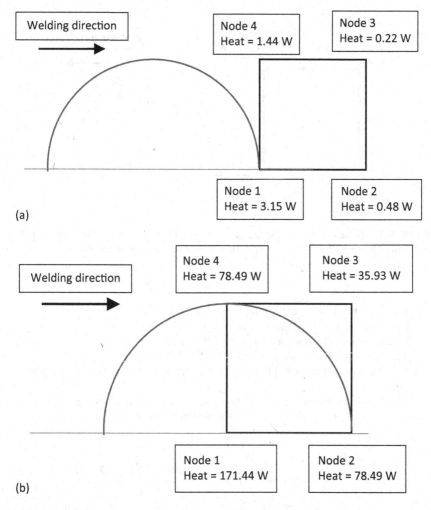

(a)

(b)

FIGURE 5.27 (a) The arc heat values at the four nodes of the element at time t = 0 s, (b) the arc heat values at the four nodes of the element at time t = 1 s. *(Continued)*

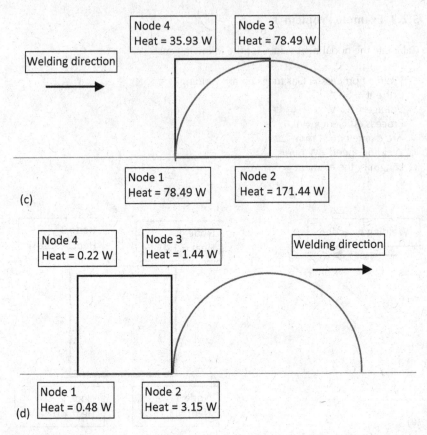

(c)

(d)

FIGURE 5.27 (Continued)   (c) the arc heat values at the four nodes of the element at time t = 2 s, and (d) the arc heat values at the four nodes of the element at time t = 3 s.

Width of the element = 2.5 mm
Distance moved by the arc for one time step = 0.5 mm
To calculate the heat values for a time interval of 1 s.
The calculated values are presented in Table 5.18.
The calculations are repeated for various time intervals and the results are given in Table 5.19.

The position of the arc for various time steps and the corresponding heat values at the four nodes of the element are presented in Figure 5.28.

TABLE 5.18
Calculation of Heat Input for t = 1 s

S. No.	x	y	N1	N2	N3	N4	q	Wx	Wy	hx	hy	F1	F2	F3	F4
1	0	0	1.0	0.0	0.0	0.0	64.96	1	1	0.5	0.25	0.90	0.0	0.0	0.0
2	0	0.25	0.9	0.0	0.0	0.1	61.98	1	4	0.5	0.25	3.10	0.0	0.0	0.34
3	0	0.5	0.8	0.0	0.0	0.2	53.85	1	2	0.5	0.25	1.20	0.0	0.0	0.30
4	0	0.75	0.7	0.0	0.0	0.3	42.60	1	4	0.5	0.25	1.66	0.0	0.0	0.71
5	0	1.0	0.6	0.0	0.0	0.4	30.68	1	2	0.5	0.25	0.51	0.0	0.0	0.34
6	0	1.25	0.5	0.0	0.0	0.5	20.12	1	4	0.5	0.25	0.56	0.0	0.0	0.56
7	0	1.5	0.4	0.0	0.0	0.6	12.02	1	2	0.5	0.25	0.13	0.0	0.0	0.20
8	0	1.75	0.3	0.0	0.0	0.7	6.53	1	4	0.5	0.25	0.11	0.0	0.0	0.25
9	0	2.0	0.2	0.0	0.0	0.8	3.23	1	2	0.5	0.25	0.02	0.0	0.0	0.07
10	0	2.25	0.1	0.0	0.0	0.9	1.46	1	4	0.5	0.25	0.01	0.0	0.0	0.07
11	0	2.5	0.0	0.0	0.0	1.0	0.60	1	1	0.5	0.25	0.0	0.0	0.0	0.01
12	0.5	0	0.9	0.1	0.0	0.0	25.44	4	1	0.5	0.25	1.27	0.14	0.0	0.0
13	0.5	0.25	0.81	0.09	0.01	0.09	24.27	4	4	0.5	0.25	4.37	0.49	0.05	0.49
14	0.5	0.5	0.72	0.08	0.02	0.18	21.09	4	2	0.5	0.25	1.69	0.19	0.05	0.42
15	0.5	0.75	0.63	0.07	0.03	0.27	16.68	4	4	0.5	0.25	2.34	0.26	0.11	1.00
16	0.5	1.0	0.54	0.06	0.04	0.36	12.02	4	2	0.5	0.25	0.72	0.08	0.05	0.48
17	0.5	1.25	0.45	0.05	0.05	0.45	7.88	4	4	0.5	0.25	0.79	0.09	0.09	0.79
18	0.5	1.5	0.36	0.04	0.06	0.54	4.71	4	2	0.5	0.25	0.19	0.02	0.03	0.28
19	0.5	1.75	0.27	0.03	0.07	0.63	2.56	4	4	0.5	0.25	0.15	0.02	0.04	0.36
20	0.5	2.0	0.18	0.02	0.08	0.72	1.27	4	2	0.5	0.25	0.03	0.0	0.01	0.10
21	0.5	2.25	0.09	0.01	0.09	0.81	0.57	4	4	0.5	0.25	0.01	0.0	0.01	0.10
22	0.5	2.5	0.0	0.0	0.1	0.9	0.23	4	1	0.5	0.25	0.0	0.0	0.0	0.01
23	1.0	0	0.8	0.2	0.0	0.0	6.85	2	1	0.5	0.25	0.15	0.04	0.0	0.0
24	1.0	0.25	0.72	0.18	0.02	0.08	6.53	2	4	0.5	0.25	0.52	0.13	0.01	0.06
25	1.0	0.5	0.64	0.16	0.04	0.16	5.68	2	2	0.5	0.25	0.20	0.05	0.01	0.05
26	1.0	0.75	0.56	0.14	0.06	0.24	4.49	2	4	0.5	0.25	0.28	0.07	0.03	0.12
27	1.0	1.0	0.48	0.12	0.08	0.32	3.23	2	2	0.5	0.25	0.09	0.02	0.01	0.06
28	1.0	1.25	0.4	0.1	0.1	0.4	2.12	2	4	0.5	0.25	0.09	0.02	0.02	0.09
29	1.0	1.5	0.32	0.08	0.12	0.48	1.27	2	2	0.5	0.25	0.02	0.01	0.01	0.03
30	1.0	1.75	0.24	0.06	0.14	0.56	0.69	2	4	0.5	0.25	0.02	0.0	0.01	0.04
31	1.0	2.0	0.16	0.04	0.16	0.64	0.34	2	2	0.5	0.25	0.0	0.0	0.0	0.01
32	1.0	2.25	0.08	0.02	0.18	0.72	0.15	2	4	0.5	0.25	0.0	0.0	0.0	0.01
33	1.0	2.5	0.0	0.0	0.2	0.8	0.06	2	1	0.5	0.25	0.0	0.0	0.0	0.0
34	1.5	0	0.7	0.3	0.0	0.0	1.27	4	1	0.5	0.25	0.05	0.02	0.0	0.0
35	1.5	0.25	0.63	0.27	0.03	0.07	1.21	4	4	0.5	0.25	0.17	0.07	0.01	0.02
36	1.5	0.5	0.56	0.24	0.06	0.14	1.05	4	2	0.5	0.25	0.07	0.03	0.01	0.02
37	1.5	0.75	0.49	0.21	0.09	0.21	0.83	4	4	0.5	0.25	0.09	0.04	0.02	0.04
38	1.5	1.0	0.42	0.18	0.12	0.28	0.60	4	2	0.5	0.25	0.03	0.01	0.01	0.02
39	1.5	1.25	0.35	0.15	0.15	0.35	0.39	4	4	0.5	0.25	0.03	0.01	0.01	0.03
40	1.5	1.5	0.28	0.12	0.18	0.42	0.23	4	2	0.5	0.25	0.01	0.0	0.0	0.01

(Continued)

TABLE 5.18 (Continued)

Calculation of Heat Input for t = 1 s

S. No.	x	y	N1	N2	N3	N4	q	Wx	Wy	hx	hy	F1	F2	F3	F4
41	1.5	1.75	0.21	0.09	0.21	0.49	0.13	4	4	0.5	0.25	0.01	0.0	0.01	0.01
42	1.5	2.0	0.14	0.06	0.24	0.56	0.06	4	2	0.5	0.25	0.0	0.0	0.0	0.0
43	1.5	2.25	0.07	0.03	0.27	0.63	0.03	4	4	0.5	0.25	0.0	0.0	0.0	0.0
44	1.5	2.5	0.0	0.0	0.3	0.7	0.01	4	1	0.5	0.25	0.0	0.0	0.0	0.0
45	2.0	0	0.6	0.4	0.0	0.0	0.16	2	1	0.5	0.25	0.0	0.0	0.0	0.0
46	2.0	0.25	0.54	0.36	0.04	0.06	0.15	2	4	0.5	0.25	0.01	0.01	0.0	0.0
47	2.0	0.5	0.48	0.32	0.08	0.12	0.13	2	2	0.5	0.25	0.0	0.0	0.0	0.0
48	2.0	0.75	0.42	0.28	0.12	0.18	0.11	2	4	0.5	0.25	0.0	0.0	0.0	0.0
49	2.0	1.0	0.36	0.24	0.16	0.24	0.08	2	2	0.5	0.25	0.0	0.0	0.0	0.0
50	2.0	1.25	0.3	0.2	0.2	0.3	0.05	2	4	0.5	0.25	0.0	0.0	0.0	0.0
51	2.0	1.5	0.24	0.16	0.24	0.36	0.03	2	2	0.5	0.25	0.0	0.0	0.0	0.0
52	2.0	1.75	0.18	0.12	0.28	0.42	0.02	2	4	0.5	0.25	0.0	0.0	0.0	0.0
53	2.0	2.0	0.12	0.08	0.32	0.48	0.01	2	2	0.5	0.25	0.0	0.0	0.0	0.0
54	2.0	2.25	0.06	0.04	0.36	0.54	0.0	2	4	0.5	0.25	0.0	0.0	0.0	0.0
55	2.0	2.5	0.0	0.0	0.4	0.6	0.0	2	1	0.5	0.25	0.0	0.0	0.0	0.0
56	2.5	0	0.5	0.5	0.0	0.0	0.01	4	1	0.5	0.25	0.0	0.0	0.0	0.0
57	2.5	0.25	0.45	0.45	0.05	0.05	0.01	4	4	0.5	0.25	0.0	0.0	0.0	0.0
58	2.5	0.5	0.4	0.4	0.1	0.1	0.01	4	2	0.5	0.25	0.0	0.0	0.0	0.0
59	2.5	0.75	0.35	0.35	0.15	0.15	0.01	4	4	0.5	0.25	0.0	0.0	0.0	0.0
60	2.5	1.0	0.3	0.3	0.2	0.2	0.01	4	2	0.5	0.25	0.0	0.0	0.0	0.0
61	2.5	1.25	0.25	0.25	0.25	0.25	0.0	4	4	0.5	0.25	0.0	0.0	0.0	0.0
62	2.5	1.5	0.2	0.2	0.3	0.3	0.0	4	2	0.5	0.25	0.0	0.0	0.0	0.0
63	2.5	1.75	0.15	0.15	0.35	0.35	0.0	4	4	0.5	0.25	0.0	0.0	0.0	0.0
64	2.5	2.0	0.1	0.1	0.4	0.4	0.0	4	2	0.5	0.25	0.0	0.0	0.0	0.0
65	2.5	2.25	0.05	0.05	0.45	0.45	0.0	4	4	0.5	0.25	0.0	0.0	0.0	0.0
66	2.5	2.5	0.0	0.0	0.5	0.5	0.0	4	1	0.5	0.25	0.0	0.0	0.0	0.0
67	3.0	0	0.4	0.6	0.0	0.0	0.0	2	1	0.5	0.25	0.0	0.0	0.0	0.0
68	3.0	0.25	0.36	0.54	0.06	0.04	0.0	2	4	0.5	0.25	0.0	0.0	0.0	0.0
69	3.0	0.5	0.32	0.48	0.12	0.08	0.0	2	2	0.5	0.25	0.0	0.0	0.0	0.0
70	3.0	0.75	0.28	0.42	0.18	0.12	0.0	2	4	0.5	0.25	0.0	0.0	0.0	0.0
71	3.0	1.0	0.24	0.36	0.24	0.16	0.0	2	2	0.5	0.25	0.0	0.0	0.0	0.0
72	3.0	1.25	0.2	0.3	0.3	0.2	0.0	2	4	0.5	0.25	0.0	0.0	0.0	0.0
73	3.0	1.5	0.16	0.24	0.36	0.24	0.0	2	2	0.5	0.25	0.0	0.0	0.0	0.0
74	3.0	1.75	0.12	0.18	0.42	0.28	0.0	2	4	0.5	0.25	0.0	0.0	0.0	0.0
75	3.0	2.0	0.08	0.12	0.48	0.32	0.0	2	2	0.5	0.25	0.0	0.0	0.0	0.0
76	3.0	2.25	0.04	0.06	0.54	0.36	0.0	2	4	0.5	0.25	0.0	0.0	0.0	0.0
77	3.0	2.5	0.0	0.0	0.6	0.4	0.0	2	1	0.5	0.25	0.0	0.0	0.0	0.0
78	3.5	0	0.3	0.7	0.0	0.0	0.0	4	1	0.5	0.25	0.0	0.0	0.0	0.0
79	3.5	0.25	0.27	0.63	0.07	0.03	0.0	4	4	0.5	0.25	0.0	0.0	0.0	0.0
80	3.5	0.5	0.24	0.56	0.14	0.06	0.0	4	2	0.5	0.25	0.0	0.0	0.0	0.0
81	3.5	0.75	0.21	0.49	0.21	0.09	0.0	4	4	0.5	0.25	0.0	0.0	0.0	0.0

(Continued)

TABLE 5.18 (*Continued*)
Calculation of Heat Input for t = 1 s

S. No.	x	y	N1	N2	N3	N4	q	Wx	Wy	hx	hy	F1	F2	F3	F4
82	3.5	1.0	0.18	0.42	0.28	0.12	0.0	4	2	0.5	0.25	0.0	0.0	0.0	0.0
83	3.5	1.25	0.15	0.35	0.35	0.15	0.0	4	4	0.5	0.25	0.0	0.0	0.0	0.0
84	3.5	1.5	0.12	0.28	0.42	0.18	0.0	4	2	0.5	0.25	0.0	0.0	0.0	0.0
85	3.5	1.75	0.09	0.21	0.49	0.21	0.0	4	4	0.5	0.25	0.0	0.0	0.0	0.0
86	3.5	2.0	0.06	0.14	0.56	0.24	0.0	4	2	0.5	0.25	0.0	0.0	0.0	0.0
87	3.5	2.25	0.03	0.07	0.63	0.27	0.0	4	4	0.5	0.25	0.0	0.0	0.0	0.0
88	3.5	2.5	0.0	0.0	0.7	0.3	0.0	4	1	0.5	0.25	0.0	0.0	0.0	0.0
89	4.0	0	0.2	0.8	0.0	0.0	0.0	2	1	0.5	0.25	0.0	0.0	0.0	0.0
90	4.0	0.25	0.18	0.72	0.08	0.02	0.0	2	4	0.5	0.25	0.0	0.0	0.0	0.0
91	4.0	0.5	0.16	0.64	0.16	0.04	0.0	2	2	0.5	0.25	0.0	0.0	0.0	0.0
92	4.0	0.75	0.14	0.56	0.24	0.06	0.0	2	4	0.5	0.25	0.0	0.0	0.0	0.0
93	4.0	1.0	0.12	0.48	0.32	0.08	0.0	2	2	0.5	0.25	0.0	0.0	0.0	0.0
94	4.0	1.25	0.1	0.4	0.4	0.1	0.0	2	4	0.5	0.25	0.0	0.0	0.0	0.0
95	4.0	1.5	0.08	0.32	0.48	0.12	-0.0	2	2	0.5	0.25	0.0	0.0	0.0	0.0
96	4.0	1.75	0.06	0.24	0.56	0.14	0.0	2	4	0.5	0.25	0.0	0.0	0.0	0.0
97	4.0	2.0	0.04	0.16	0.64	0.16	0.0	2	2	0.5	0.25	0.0	0.0	0.0	0.0
98	4.0	2.25	0.02	0.08	0.72	0.18	0.0	2	4	0.5	0.25	0.0	0.0	0.0	0.0
99	4.0	2.5	0.0	0.0	0.8	0.2	0.0	2	1	0.5	0.25	0.0	0.0	0.0	0.0
100	4.5	0	0.1	0.9	0.0	0.0	0.0	4	1	0.5	0.25	0.0	0.0	0.0	0.0
101	4.5	0.25	0.09	0.81	0.09	0.01	0.0	4	4	0.5	0.25	0.0	0.0	0.0	0.0
102	4.5	0.5	0.08	0.72	0.18	0.02	0.0	4	2	0.5	0.25	0.0	0.0	0.0	0.0
103	4.5	0.75	0.07	0.63	0.27	0.03	0.0	4	4	0.5	0.25	0.0	0.0	0.0	0.0
104	4.5	1.0	0.06	0.54	0.36	0.04	0.0	4	2	0.5	0.25	0.0	0.0	0.0	0.0
105	4.5	1.25	0.05	0.45	0.45	0.05	0.0	4	4	0.5	0.25	0.0	0.0	0.0	0.0
106	4.5	1.5	0.04	0.36	0.54	0.06	0.0	4	2	0.5	0.25	0.0	0.0	0.0	0.0
107	4.5	1.75	0.03	0.27	0.63	0.07	0.0	4	4	0.5	0.25	0.0	0.0	0.0	0.0
108	4.5	2.0	0.02	0.18	0.72	0.08	0.0	4	2	0.5	0.25	0.0	0.0	0.0	0.0
109	4.5	2.25	0.01	0.09	0.81	0.09	0.0	4	4	0.5	0.25	0.0	0.0	0.0	0.0
110	4.5	2.5	0.0	0.0	0.9	0.1	0.0	4	1	0.5	0.25	0.0	0.0	0.0	0.0
111	5.0	0	0.0	1.0	0.0	0.0	0.0	1	1	0.5	0.25	0.0	0.0	0.0	0.0
112	5.0	0.25	0.0	0.9	0.1	0.0	0.0	1	4	0.5	0.25	0.0	0.0	0.0	0.0
113	5.0	0.5	0.0	0.8	0.2	0.0	0.0	1	2	0.5	0.25	0.0	0.0	0.0	0.0
114	5.0	0.75	0.0	0.7	0.3	0.0	0.0	1	4	0.5	0.25	0.0	0.0	0.0	0.0
115	5.0	1.0	0.0	0.6	0.4	0.0	0.0	1	2	0.5	0.25	0.0	0.0	0.0	0.0
116	5.0	1.25	0.0	0.5	0.5	0.0	0.0	1	4	0.5	0.25	0.0	0.0	0.0	0.0
117	5.0	1.5	0.0	0.4	0.6	0.0	0.0	1	2	0.5	0.25	0.0	0.0	0.0	0.0
118	5.0	1.75	0.0	0.3	0.7	0.0	0.0	1	4	0.5	0.25	0.0	0.0	0.0	0.0
119	5.0	2.0	0.0	0.2	0.8	0.0	0.0	1	2	0.5	0.25	0.0	0.0	0.0	0.0
120	5.0	2.25	0.0	0.1	0.9	0.0	0.0	1	4	0.5	0.25	0.0	0.0	0.0	0.0
121	5.0	2.5	0.0	0.0	1.0	0.0	0.0	1	1	0.5	0.25	0.0	0.0	0.0	0.0
Total												21.60	1.84	0.64	7.54

TABLE 5.19

Arc Heat Calculated for Various Time Steps

Time, s	Heat Input, watts			
	Node 1	Node 2	Node 3	Node 4
0	1.48	0.07	0.03	0.52
0.5	6.63	0.44	0.15	2.32
1.0	21.60	1.84	0.64	7.54
1.5	51.66	5.77	2.01	18.04
2.0	92.58	13.92	4.86	32.32
2.5	128.50	27.07	9.45	44.87
3.0	145.11	44.44	15.52	50.67
3.5	141.59	64.33	22.46	49.44
4.0	126.17	85.24	29.76	44.05
4.5	106.25	106.25	37.10	37.10
5.0	85.24	126.17	44.05	29.76
5.5	64.33	141.59	49.44	22.46
6.0	44.44	145.11	50.67	15.52
6.5	27.07	128.50	44.87	9.45
7.0	13.92	92.58	32.32	4.86
7.5	5.77	51.66	18.04	2.01
8.0	1.84	21.60	7.54	0.64
8.5	0.44	6.63	2.32	0.15
9.0	0.07	1.48	0.52	0.03

5.2.4 Example Problem 4

Calculate the nodal heat values for the following case.

Welding process = Gas metal arc welding
Current = 70 A
Voltage = 16 V
Process efficiency = 0.786
Arc diameter = 8 mm
Welding speed = 4 mm/s
Length of the element = 2 mm
Width of the element = 2 mm
Distance moved by the arc for each time step = 0.5 mm

To calculate the heat values for a time interval of 2 s.

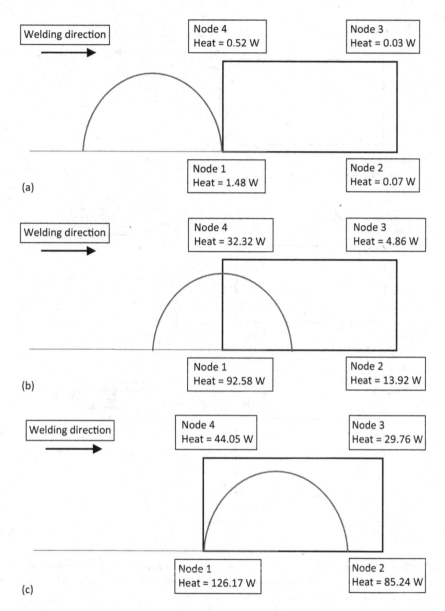

FIGURE 5.28 (a) The arc heat values at the four nodes of the element at time t = 0 s; (b) the arc heat values at the four nodes of the element at time t = 2 s; (c) the arc heat values at the four nodes of the element at time t = 4 s. (*Continued*)

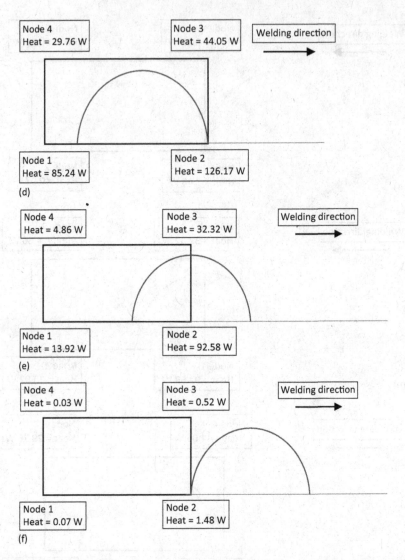

| Node 4
Heat = 29.76 W | Node 3
Heat = 44.05 W | Welding direction → |
| Node 1
Heat = 85.24 W | Node 2
Heat = 126.17 W | |

(d)

| Node 4
Heat = 4.86 W | Node 3
Heat = 32.32 W | Welding direction → |
| Node 1
Heat = 13.92 W | Node 2
Heat = 92.58 W | |

(e)

| Node 4
Heat = 0.03 W | Node 3
Heat = 0.52 W | Welding direction → |
| Node 1
Heat = 0.07 W | Node 2
Heat = 1.48 W | |

(f)

FIGURE 5.28 (*Continued*)   (d) the arc heat values at the four nodes of the element at time t = 5 s, (e) the arc heat values at the four nodes of the element at time t = 7 s, and (f) the arc heat values at the four nodes of the element at time t = 9 s.

In this case, there are two elements in the width direction as shown in Figure 5.29. The element 1 has y coordinate from 0 to 2 mm whereas element 2 has y coordinate from 2 to 4 mm. For evaluating the heat input, the following generalized shape functions are to be employed.

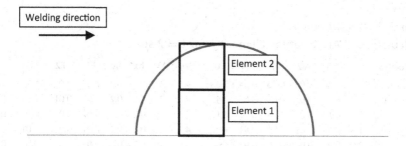

FIGURE 5.29 The position of the arc and the elements at time t = 2 s.

$$N1 = (1 - x/L)(W2 - y)/(W2 - W1)$$

$$N2 = (x/L)(W2 - y)/(W2 - W1)$$

$$N3 = (x/L)(y - W1)/(W2 - W1)$$

$$N4 = (1 - x/L)(y - W1)/(W2 - W1)$$

Table 5.20 shows the calculated values for the first element having a y coordinate from 0 to 2 mm. The calculated arc heat values for various time intervals are presented in Table 5.21. Tables 5.22 and 5.23 show the corresponding results for the second element with y coordinate ranging from 2 to 4 mm.

TABLE 5.20

Calculation of Arc Heat for Element 1 for t = 2 sec

S. No.	x	y	N1	N2	N3	N4	q	Wx	Wy	hx	hy	F1	F2	F3	F4
1	0	0	1.0	0.0	0.0	0.0	2.62	1	1	0.2	0.2	0.01	0.0	0.0	0.0
2	0	0.2	0.9	0.0	0.0	0.1	2.60	1	4	0.2	0.2	0.04	0.0	0.0	0.0
3	0	0.4	0.8	0.0	0.0	0.2	2.54	1	2	0.2	0.2	0.02	0.0	0.0	0.0
4	0	0.6	0.7	0.0	0.0	0.3	2.45	1	4	0.2	0.2	0.03	0.0	0.0	0.01
5	0	0.8	0.6	0.0	0.0	0.4	2.32	1	2	0.2	0.2	0.01	0.0	0.0	0.01
6	0	1.0	0.5	0.0	0.0	0.5	2.17	1	4	0.2	0.2	0.02	0.0	0.0	0.02
7	0	1.2	0.4	0.0	0.0	0.6	2.00	1	2	0.2	0.2	0.01	0.0	0.0	0.01
8	0	1.4	0.3	0.0	0.0	0.7	1.81	1	4	0.2	0.2	0.01	0.0	0.0	0.02
9	0	1.6	0.2	0.0	0.0	0.8	1.62	1	2	0.2	0.2	0.0	0.0	0.0	0.01
10	0	1.8	0.1	0.0	0.0	0.9	1.42	1	4	0.2	0.2	0.0	0.0	0.0	0.02
11	0	2.0	0.0	0.0	0.0	1.0	1.24	1	1	0.2	0.2	0.0	0.0	0.0	0.01
12	0.2	0	0.9	0.1	0.0	0.0	3.50	4	1	0.2	0.2	0.06	0.01	0.0	0.0
13	0.2	0.2	0.81	0.09	0.01	0.09	3.48	4	4	0.2	0.2	0.20	0.02	0.0	0.02
14	0.2	0.4	0.72	0.08	0.02	0.18	3.40	4	2	0.2	0.2	0.09	0.01	0.0	0.02
15	0.2	0.6	0.63	0.07	0.03	0.27	3.28	4	4	0.2	0.2	0.15	0.02	0.01	0.06
16	0.2	0.8	0.54	0.06	0.04	0.36	3.11	4	2	0.2	0.2	0.06	0.01	0.0	0.04
17	0.2	1.0	0.45	0.05	0.05	0.45	2.91	4	4	0.2	0.2	0.09	0.01	0.01	0.09

(Continued)

TABLE 5.20 (*Continued*)

Calculation of Arc Heat for Element 1 for t = 2 sec

S. No.	x	y	N1	N2	N3	N4	q	Wx	Wy	hx	hy	F1	F2	F3	F4
18	0.2	1.2	0.36	0.04	0.06	0.54	2.68	4	2	0.2	0.2	0.03	0.0	0.01	0.05
19	0.2	1.4	0.27	0.03	0.07	0.63	2.43	4	4	0.2	0.2	0.05	0.01	0.01	0.11
20	0.2	1.6	0.18	0.02	0.08	0.72	2.17	4	2	0.2	0.2	0.01	0.0	0.01	0.06
21	0.2	1.8	0.09	0.01	0.09	0.81	1.91	4	4	0.2	0.2	0.01	0.0	0.01	0.11
22	0.2	2.0	0.0	0.0	0.1	0.9	1.66	4	1	0.2	0.2	0.0	0.0	0.0	0.03
23	0.4	0	0.8	0.2	0.0	0.0	4.63	2	1	0.2	0.2	0.03	0.01	0.0	0.0
24	0.4	0.2	0.72	0.18	0.02	0.08	4.59	2	4	0.2	0.2	0.12	0.03	0.0	0.01
25	0.4	0.4	0.64	0.16	0.04	0.16	4.49	2	2	0.2	0.2	0.05	0.01	0.0	0.01
26	0.4	0.6	0.56	0.14	0.06	0.24	4.32	2	4	0.2	0.2	0.09	0.02	0.01	0.04
27	0.4	0.8	0.48	0.12	0.08	0.32	4.10	2	2	0.2	0.2	0.04	0.01	0.01	0.02
28	0.4	1.0	0.4	0.1	0.1	0.4	3.83	2	4	0.2	0.2	0.05	0.01	0.01	0.05
29	0.4	1.2	0.32	0.08	0.12	0.48	3.53	2	2	0.2	0.2	0.02	0.01	0.01	0.03
30	0.4	1.4	0.24	0.06	0.14	0.56	3.20	2	4	0.2	0.2	0.03	0.01	0.02	0.06
31	0.4	1.6	0.16	0.04	0.16	0.64	2.86	2	2	0.2	0.2	0.01	0.0	0.01	0.03
32	0.4	1.8	0.08	0.02	0.18	0.72	2.52	2	4	0.2	0.2	0.01	0.0	0.02	0.06
33	0.4	2.0	0.0	0.0	0.2	0.8	2.18	2	1	0.2	0.2	0.0	0.0	0.0	0.02
34	0.6	0	0.7	0.3	0.0	0.0	6.01	4	1	0.2	0.2	0.07	0.03	0.0	0.0
35	0.6	0.2	0.63	0.27	0.03	0.07	5.97	4	4	0.2	0.2	0.27	0.11	0.01	0.03
36	0.6	0.4	0.56	0.24	0.06	0.14	5.84	4	2	0.2	0.2	0.12	0.05	0.01	0.03
37	0.6	0.6	0.49	0.21	0.09	0.21	5.62	4	4	0.2	0.2	0.20	0.08	0.04	0.08
38	0.6	0.8	0.42	0.18	0.12	0.28	5.33	4	2	0.2	0.2	0.08	0.03	0.02	0.05
39	0.6	1.0	0.35	0.15	0.15	0.35	4.99	4	4	0.2	0.2	0.12	0.05	0.05	0.12
40	0.6	1.2	0.28	0.12	0.18	0.42	4.59	4	2	0.2	0.2	0.05	0.02	0.03	0.07
41	0.6	1.4	0.21	0.09	0.21	0.49	4.16	4	4	0.2	0.2	0.06	0.03	0.06	0.15
42	0.6	1.6	0.14	0.06	0.24	0.56	3.72	4	2	0.2	0.2	0.02	0.01	0.03	0.07
43	0.6	1.8	0.07	0.03	0.27	0.63	3.28	4	4	0.2	0.2	0.02	0.01	0.06	0.15
44	0.6	2.0	0.0	0.0	0.3	0.7	2.84	4	1	0.2	0.2	0.0	0.0	0.02	0.04
45	0.8	0	0.6	0.4	0.0	0.0	7.70	2	1	0.2	0.2	0.04	0.03	0.0	0.0
46	0.8	0.2	0.54	0.36	0.04	0.06	7.65	2	4	0.2	0.2	0.15	0.10	0.01	0.02
47	0.8	0.4	0.48	0.32	0.08	0.12	7.48	2	2	0.2	0.2	0.06	0.04	0.01	0.02
48	0.8	0.6	0.42	0.28	0.12	0.18	7.20	2	4	0.2	0.2	0.11	0.07	0.03	0.05
49	0.8	0.8	0.36	0.24	0.16	0.24	6.83	2	2	0.2	0.2	0.04	0.03	0.02	0.03
50	0.8	1.0	0.3	0.2	0.2	0.3	6.39	2	4	0.2	0.2	0.07	0.05	0.05	0.07
51	0.8	1.2	0.24	0.16	0.24	0.36	5.88	2	2	0.2	0.2	0.03	0.02	0.03	0.04
52	0.8	1.4	0.18	0.12	0.28	0.42	5.33	2	4	0.2	0.2	0.03	0.02	0.05	0.08
53	0.8	1.6	0.12	0.08	0.32	0.48	4.77	2	2	0.2	0.2	0.01	0.01	0.03	0.04
54	0.8	1.8	0.06	0.04	0.36	0.54	4.20	2	4	0.2	0.2	0.01	0.01	0.05	0.08
55	0.8	2.0	0.0	0.0	0.4	0.6	3.64	2	1	0.2	0.2	0.0	0.0	0.01	0.02
56	1.0	0	0.5	0.5	0.0	0.0	9.72	4	1	0.2	0.2	0.09	0.09	0.0	0.0
57	1.0	0.2	0.45	0.45	0.05	0.05	9.65	4	4	0.2	0.2	0.31	0.31	0.03	0.03
58	1.0	0.4	0.4	0.4	0.1	0.1	9.43	4	2	0.2	0.2	0.13	0.13	0.03	0.03

(Continued)

TABLE 5.20 (*Continued*)
Calculation of Arc Heat for Element 1 for t = 2 sec

S. No.	x	y	N1	N2	N3	N4	q	Wx	Wy	hx	hy	F1	F2	F3	F4
59	1.0	0.6	0.35	0.35	0.15	0.15	9.08	4	4	0.2	0.2	0.23	0.23	0.10	0.10
60	1.0	0.8	0.3	0.3	0.2	0.2	8.62	4	2	0.2	0.2	0.09	0.09	0.06	0.06
61	1.0	1.0	0.25	0.25	0.25	0.25	8.06	4	4	0.2	0.2	0.14	0.14	0.14	0.14
62	1.0	1.2	0.2	0.2	0.3	0.3	7.42	4	2	0.2	0.2	0.05	0.05	0.08	0.08
63	1.0	1.4	0.15	0.15	0.35	0.35	6.73	4	4	0.2	0.2	0.07	0.07	0.17	0.17
64	1.0	1.6	0.1	0.1	0.4	0.4	6.01	4	2	0.2	0.2	0.02	0.02	0.09	0.09
65	1.0	1.8	0.05	0.05	0.45	0.45	5.29	4	4	0.2	0.2	0.02	0.02	0.17	0.17
66	1.0	2.0	0.0	0.0	0.5	0.5	4.59	4	1	0.2	0.2	0.0	0.0	0.04	0.04
67	1.2	0	0.4	0.6	0.0	0.0	12.08	2	1	0.2	0.2	0.04	0.06	0.0	0.0
68	1.2	0.2	0.36	0.54	0.06	0.04	11.99	2	4	0.2	0.2	0.15	0.23	0.03	0.02
69	1.2	0.4	0.32	0.48	0.12	0.08	11.72	2	2	0.2	0.2	0.07	0.10	0.03	0.02
70	1.2	0.6	0.28	0.42	0.18	0.12	11.29	2	4	0.2	0.2	0.11	0.17	0.07	0.05
71	1.2	0.8	0.24	0.36	0.24	0.16	10.71	2	2	0.2	0.2	0.05	0.07	0.05	0.03
72	1.2	1.0	0.2	0.3	0.3	0.2	10.01	2	4	0.2	0.2	0.07	0.11	0.11	0.07
73	1.2	1.2	0.16	0.24	0.36	0.24	9.22	2	2	0.2	0.2	0.03	0.04	0.06	0.04
74	1.2	1.4	0.12	0.18	0.42	0.28	8.37	2	4	0.2	0.2	0.04	0.05	0.12	0.08
75	1.2	1.6	0.08	0.12	0.48	0.32	7.48	2	2	0.2	0.2	0.01	0.02	0.06	0.04
76	1.2	1.8	0.04	0.06	0.54	0.36	6.58	2	4	0.2	0.2	0.01	0.01	0.13	0.08
77	1.2	2.0	0.0	0.0	0.6	0.4	5.71	2	1	0.2	0.2	0.0	0.0	0.03	0.02
78	1.4	0	0.3	0.7	0.0	0.0	14.79	4	1	0.2	0.2	0.08	0.18	0.0	0.0
79	1.4	0.2	0.27	0.63	0.07	0.03	14.68	4	4	0.2	0.2	0.28	0.66	0.07	0.03
80	1.4	0.4	0.24	0.56	0.14	0.06	14.35	4	2	0.2	0.2	0.12	0.29	0.07	0.03
81	1.4	0.6	0.21	0.49	0.21	0.09	13.83	4	4	0.2	0.2	0.21	0.48	0.21	0.09
82	1.4	0.8	0.18	0.42	0.28	0.12	13.12	4	2	0.2	0.2	0.08	0.20	0.13	0.06
83	1.4	1.0	0.15	0.35	0.35	0.15	12.26	4	4	0.2	0.2	0.13	0.31	0.31	0.13
84	1.4	1.2	0.12	0.28	0.42	0.18	11.29	4	2	0.2	0.2	0.05	0.11	0.17	0.07
85	1.4	1.4	0.09	0.21	0.49	0.21	10.24	4	4	0.2	0.2	0.07	0.15	0.36	0.15
86	1.4	1.6	0.06	0.14	0.56	0.24	9.15	4	2	0.2	0.2	0.02	0.05	0.18	0.08
87	1.4	1.8	0.03	0.07	0.63	0.27	8.06	4	4	0.2	0.2	0.02	0.04	0.36	0.15
88	1.4	2.0	0.0	0.0	0.7	0.3	6.99	4	1	0.2	0.2	0.0	0.0	0.09	0.04
89	1.6	0	0.2	0.8	0.0	0.0	17.84	2	1	0.2	0.2	0.03	0.13	0.0	0.0
90	1.6	0.2	0.18	0.72	0.08	0.02	17.71	2	4	0.2	0.2	0.11	0.45	0.05	0.01
91	1.6	0.4	0.16	0.64	0.16	0.04	17.32	2	2	0.2	0.2	0.05	0.20	0.05	0.01
92	1.6	0.6	0.14	0.56	0.24	0.06	16.68	2	4	0.2	0.2	0.08	0.33	0.14	0.04
93	1.6	0.8	0.12	0.48	0.32	0.08	15.82	2	2	0.2	0.2	0.03	0.14	0.09	0.02
94	1.6	1.0	0.1	0.4	0.4	0.1	14.79	2	4	0.2	0.2	0.05	0.21	0.21	0.05
95	1.6	1.2	0.08	0.32	0.48	0.12	13.62	2	2	0.2	0.2	0.02	0.08	0.12	0.03
96	1.6	1.4	0.06	0.24	0.56	0.14	12.36	2	4	0.2	0.2	0.03	0.11	0.25	0.06
97	1.6	1.6	0.04	0.16	0.64	0.16	11.04	2	2	0.2	0.2	0.01	0.03	0.13	0.03
98	1.6	1.8	0.02	0.08	0.72	0.18	9.72	2	4	0.2	0.2	0.01	0.03	0.25	0.06
99	1.6	2.0	0.0	0.0	0.8	0.2	8.43	2	1	0.2	0.2	0.0	0.0	0.06	0.01

(Continued)

TABLE 5.20 (*Continued*)
Calculation of Arc Heat for Element 1 for t = 2 sec

S. No.	x	y	N1	N2	N3	N4	q	Wx	Wy	hx	hy	F1	F2	F3	F4
100	1.8	0	0.1	0.9	0.0	0.0	21.20	4	1	0.2	0.2	0.04	0.34	0.0	0.0
101	1.8	0.2	0.09	0.81	0.09	0.01	21.04	4	4	0.2	0.2	0.13	1.21	0.13	0.01
102	1.8	0.4	0.08	0.72	0.18	0.02	20.58	4	2	0.2	0.2	0.06	0.53	0.13	0.01
103	1.8	0.6	0.07	0.63	0.27	0.03	19.82	4	4	0.2	0.2	0.10	0.89	0.38	0.04
104	1.8	0.8	0.06	0.54	0.36	0.04	18.80	4	2	0.2	0.2	0.04	0.36	0.24	0.03
105	1.8	1.0	0.05	0.45	0.45	0.05	17.58	4	4	0.2	0.2	0.06	0.56	0.56	0.06
106	1.8	1.2	0.04	0.36	0.54	0.06	16.18	4	2	0.2	0.2	0.02	0.21	0.31	0.03
107	1.8	1.4	0.03	0.27	0.63	0.07	14.68	4	4	0.2	0.2	0.03	0.28	0.66	0.07
108	1.8	1.6	0.02	0.18	0.72	0.08	13.12	4	2	0.2	0.2	0.01	0.08	0.34	0.04
109	1.8	1.8	0.01	0.09	0.81	0.09	11.55	4	4	0.2	0.2	0.01	0.07	0.67	0.07
110	1.8	2.0	0.0	0.0	0.9	0.1	10.01	4	1	0.2	0.2	0.0	0.0	0.16	0.02
111	2.0	0	0.0	1.0	0.0	0.0	24.82	1	1	0.2	0.2	0.0	0.11	0.0	0.0
112	2.0	0.2	0.0	0.9	0.1	0.0	24.63	1	4	0.2	0.2	0.0	0.39	0.04	0.0
113	2.0	0.4	0.0	0.8	0.2	0.0	24.08	1	2	0.2	0.2	0.0	0.17	0.04	0.0
114	2.0	0.6	0.0	0.7	0.3	0.0	23.20	1	4	0.2	0.2	0.0	0.29	0.12	0.0
115	2.0	0.8	0.0	0.6	0.4	0.0	22.01	1	2	0.2	0.2	0.0	0.12	0.08	0.0
116	2.0	1.0	0.0	0.5	0.5	0.0	20.58	1	4	0.2	0.2	0.0	0.18	0.18	0.0
117	2.0	1.2	0.0	0.4	0.6	0.0	18.95	1	2	0.2	0.2	0.0	0.07	0.10	0.0
118	2.0	1.4	0.0	0.3	0.7	0.0	17.19	1	4	0.2	0.2	0.0	0.09	0.21	0.0
119	2.0	1.6	0.0	0.2	0.8	0.0	15.36	1	2	0.2	0.2	0.0	0.03	0.11	0.0
120	2.0	1.8	0.0	0.1	0.9	0.0	13.52	1	4	0.2	0.2	0.0	0.02	0.22	0.0
121	2.0	2.0	0.0	0.0	1.0	0.0	11.72	1	1	0.2	0.2	0.0	0.0	0.05	0.0
Total												6.61	13.17	10.40	5.21

Note: x varies from 0 to 2 mm and y varies from 0 to 2 mm.

TABLE 5.21
Arc Heat Values Calculated for Various Time Steps for Element 1

	Heat Input, watts			
Time, s	Node 1	Node 2	Node 3	Node 4
0	1.01	0.34	0.27	0.80
0.125	2.19	0.81	0.64	1.73
0.25	4.36	1.78	1.40	3.44
0.375	7.92	3.57	2.82	6.25
0.5	13.17	6.61	5.21	10.40
0.625	20.05	11.22	8.85	15.83
0.75	27.95	17.50	13.81	22.06

(*Continued*)

TABLE 5.21 (*Continued*)
Arc Heat Values Calculated for Various Time Steps for Element 1

	Heat Input, watts			
Time, s	Node 1	Node 2	Node 3	Node 4
0.875	35.69	25.06	19.78	28.16
1.0	41.76	32.95	26.00	32.95
1.125	44.79	39.78	31.39	35.35
1.25	44.07	44.07	34.78	34.78
1.375	39.78	44.79	35.35	31.39
1.5	32.95	41.76	32.95	26.00
1.625	25.06	35.69	28.16	19.78
1.75	17.50	27.95	22.06	13.81
1.875	11.22	20.05	15.83	8.85
2.0	6.61	13.17	10.40	5.21
2.125	3.57	7.92	6.25	2.82
2.25	1.78	4.36	3.44	1.40
2.375	0.81	2.19	1.73	0.64
2.5	0.34	1.01	0.80	0.27

TABLE 5.22
Calculation of Arc Heat for Element 2 for t = 2 s

S. No.	x	y	N1	N2	N3	N4	q	Wx	Wy	hx	hy	F1	F2	F3	F4
1	0	2.0	1.0	0.0	0.0	0.0	1.24	1	1	0.2	0.2	0.01	0.0	0.0	0.0
2	0	2.2	0.9	0.0	0.0	0.1	1.06	1	4	0.2	0.2	0.02	0.0	0.0	0.0
3	0	2.4	0.8	0.0	0.0	0.2	0.89	1	2	0.2	0.2	0.01	0.0	0.0	0.0
4	0	2.6	0.7	0.0	0.0	0.3	0.74	1	4	0.2	0.2	0.01	0.0	0.0	0.0
5	0	2.8	0.6	0.0	0.0	0.4	0.60	1	2	0.2	0.2	0.0	0.0	0.0	0.0
6	0	3.0	0.5	0.0	0.0	0.5	0.48	1	4	0.2	0.2	0.0	0.0	0.0	0.0
7	0	3.2	0.4	0.0	0.0	0.6	0.38	1	2	0.2	0.2	0.0	0.0	0.0	0.0
8	0	3.4	0.3	0.0	0.0	0.7	0.30	1	4	0.2	0.2	0.0	0.0	0.0	0.0
9	0	3.6	0.2	0.0	0.0	0.8	0.23	1	2	0.2	0.2	0.0	0.0	0.0	0.0
10	0	3.8	0.1	0.0	0.0	0.9	0.17	1	4	0.2	0.2	0.0	0.0	0.0	0.0
11	0	4.0	0.0	0.0	0.0	1.0	0.13	1	1	0.2	0.2	0.0	0.0	0.0	0.0
12	0.2	2.0	0.9	0.1	0.0	0.0	1.66	4	1	0.2	0.2	0.03	0.0	0.0	0.0
13	0.2	2.2	0.81	0.09	0.01	0.09	1.41	4	4	0.2	0.2	0.08	0.01	0.0	0.01
14	0.2	2.4	0.72	0.08	0.02	0.18	1.19	4	2	0.2	0.2	0.03	0.0	0.0	0.01
15	0.2	2.6	0.63	0.07	0.03	0.27	0.99	4	4	0.2	0.2	0.04	0.0	0.0	0.02
16	0.2	2.8	0.54	0.06	0.04	0.36	0.81	4	2	0.2	0.2	0.02	0.0	0.0	0.01
17	0.2	3.0	0.45	0.05	0.05	0.45	0.65	4	4	0.2	0.2	0.02	0.0	0.0	0.02

(Continued)

TABLE 5.22 (*Continued*)
Calculation of Arc Heat for Element 2 for t = 2 s

S. No.	x	y	N1	N2	N3	N4	q	Wx	Wy	hx	hy	F1	F2	F3	F4
18	0.2	3.2	0.36	0.04	0.06	0.54	0.51	4	2	0.2	0.2	0.01	0.0	0.0	0.01
19	0.2	3.4	0.27	0.03	0.07	0.63	0.40	4	4	0.2	0.2	0.01	0.0	0.0	0.02
20	0.2	3.6	0.18	0.02	0.08	0.72	0.31	4	2	0.2	0.2	0.0	0.0	0.0	0.01
21	0.2	3.8	0.09	0.01	0.09	0.81	0.23	4	4	0.2	0.2	0.0	0.0	0.0	0.01
22	0.2	4.0	0.0	0.0	0.1	0.9	0.17	4	1	0.2	0.2	0.0	0.0	0.0	0.0
23	0.4	2.0	0.8	0.2	0.0	0.0	2.18	2	1	0.2	0.2	0.02	0.0	0.0	0.0
24	0.4	2.2	0.72	0.18	0.02	0.08	1.87	2	4	0.2	0.2	0.05	0.01	0.0	0.01
25	0.4	2.4	0.64	0.16	0.04	0.16	1.57	2	2	0.2	0.2	0.02	0.0	0.0	0.0
26	0.4	2.6	0.56	0.14	0.06	0.24	1.30	2	4	0.2	0.2	0.03	0.01	0.0	0.01
27	0.4	2.8	0.48	0.12	0.08	0.32	1.06	2	2	0.2	0.2	0.01	0.0	0.0	0.01
28	0.4	3.0	0.4	0.1	0.1	0.4	0.86	2	4	0.2	0.2	0.01	0.0	0.0	0.01
29	0.4	3.2	0.32	0.08	0.12	0.48	0.68	2	2	0.2	0.2	0.0	0.0	0.0	0.01
30	0.4	3.4	0.24	0.06	0.14	0.56	0.53	2	4	0.2	0.2	0.0	0.0	0.0	0.01
31	0.4	3.6	0.16	0.04	0.16	0.64	0.41	2	2	0.2	0.2	0.0	0.0	0.0	0.0
32	0.4	3.8	0.08	0.02	0.18	0.72	0.31	2	4	0.2	0.2	0.0	0.0	0.0	0.01
33	0.4	4.0	0.0	0.0	0.2	0.8	0.23	2	1	0.2	0.2	0.0	0.0	0.0	0.0
34	0.6	2.0	0.7	0.3	0.0	0.0	2.84	4	1	0.2	0.2	0.04	0.02	0.0	0.0
35	0.6	2.2	0.63	0.27	0.03	0.07	2.43	4	4	0.2	0.2	0.11	0.05	0.01	0.01
36	0.6	2.4	0.56	0.24	0.06	0.14	2.04	4	2	0.2	0.2	0.04	0.02	0.0	0.01
37	0.6	2.6	0.49	0.21	0.09	0.21	1.69	4	4	0.2	0.2	0.06	0.03	0.01	0.03
38	0.6	2.8	0.42	0.18	0.12	0.28	1.38	4	2	0.2	0.2	0.02	0.01	0.01	0.01
39	0.6	3.0	0.35	0.15	0.15	0.35	1.11	4	4	0.2	0.2	0.03	0.01	0.01	0.03
40	0.6	3.2	0.28	0.12	0.18	0.42	0.88	4	2	0.2	0.2	0.01	0.0	0.01	0.01
41	0.6	3.4	0.21	0.09	0.21	0.49	0.69	4	4	0.2	0.2	0.01	0.0	0.01	0.02
42	0.6	3.6	0.14	0.06	0.24	0.56	0.53	4	2	0.2	0.2	0.0	0.0	0.0	0.01
43	0.6	3.8	0.07	0.03	0.27	0.63	0.40	4	4	0.2	0.2	0.0	0.0	0.01	0.02
44	0.6	4.0	0.0	0.0	0.3	0.7	0.30	4	1	0.2	0.2	0.0	0.0	0.0	0.0
45	0.8	2.0	0.6	0.4	0.0	0.0	3.64	2	1	0.2	0.2	0.02	0.01	0.0	0.0
46	0.8	2.2	0.54	0.36	0.04	0.06	3.11	2	4	0.2	0.2	0.06	0.04	0.0	0.01
47	0.8	2.4	0.48	0.32	0.08	0.12	2.62	2	2	0.2	0.2	0.02	0.01	0.0	0.01
48	0.8	2.6	0.42	0.28	0.12	0.18	2.17	2	4	0.2	0.2	0.03	0.02	0.01	0.01
49	0.8	2.8	0.36	0.24	0.16	0.24	1.77	2	2	0.2	0.2	0.01	0.01	0.01	0.01
50	0.8	3.0	0.3	0.2	0.2	0.3	1.42	2	4	0.2	0.2	0.02	0.01	0.01	0.02
51	0.8	3.2	0.24	0.16	0.24	0.36	1.13	2	2	0.2	0.2	0.0	0.0	0.0	0.01
52	0.8	3.4	0.18	0.12	0.28	0.42	0.88	2	4	0.2	0.2	0.01	0.0	0.01	0.01
53	0.8	3.6	0.12	0.08	0.32	0.48	0.68	2	2	0.2	0.2	0.0	0.0	0.0	0.01
54	0.8	3.8	0.06	0.04	0.36	0.54	0.51	2	4	0.2	0.2	0.0	0.0	0.01	0.01
55	0.8	4.0	0.0	0.0	0.4	0.6	0.38	2	1	0.2	0.2	0.0	0.0	0.0	0.0

(Continued)

TABLE 5.22 (*Continued*)
Calculation of Arc Heat for Element 2 for t = 2 s

S. No.	x	y	N1	N2	N3	N4	q	Wx	Wy	hx	hy	F1	F2	F3	F4
56	1.0	2.0	0.5	0.5	0.0	0.0	4.59	4	1	0.2	0.2	0.04	0.04	0.0	0.0
57	1.0	2.2	0.45	0.45	0.05	0.05	3.92	4	4	0.2	0.2	0.13	0.13	0.01	0.01
58	1.0	2.4	0.4	0.4	0.1	0.1	3.30	4	2	0.2	0.2	0.05	0.05	0.01	0.01
59	1.0	2.6	0.35	0.35	0.15	0.15	2.74	4	4	0.2	0.2	0.07	0.07	0.03	0.03
60	1.0	2.8	0.3	0.3	0.2	0.2	2.23	4	2	0.2	0.2	0.02	0.02	0.02	0.02
61	1.0	3.0	0.25	0.25	0.25	0.25	1.80	4	4	0.2	0.2	0.03	0.03	0.03	0.03
62	1.0	3.2	0.2	0.2	0.3	0.3	1.42	4	2	0.2	0.2	0.01	0.01	0.02	0.02
63	1.0	3.4	0.15	0.15	0.35	0.35	1.11	4	4	0.2	0.2	0.01	0.01	0.03	0.03
64	1.0	3.6	0.1	0.1	0.4	0.4	0.86	4	2	0.2	0.2	0.0	0.0	0.01	0.01
65	1.0	3.8	0.05	0.05	0.45	0.45	0.65	4	4	0.2	0.2	0.0	0.0	0.02	0.02
66	1.0	4.0	0.0	0.0	0.5	0.5	0.48	4	1	0.2	0.2	0.0	0.0	0.0	0.0
67	1.2	2.0	0.4	0.6	0.0	0.0	5.71	2	1	0.2	0.2	0.02	0.03	0.0	0.0
68	1.2	2.2	0.36	0.54	0.06	0.04	4.87	2	4	0.2	0.2	0.06	0.09	0.01	0.01
69	1.2	2.4	0.32	0.48	0.12	0.08	4.10	2	2	0.2	0.2	0.02	0.04	0.01	0.01
70	1.2	2.6	0.28	0.42	0.18	0.12	3.40	2	4	0.2	0.2	0.03	0.05	0.02	0.01
71	1.2	2.8	0.24	0.36	0.24	0.16	2.78	2	2	0.2	0.2	0.01	0.02	0.01	0.01
72	1.2	3.0	0.2	0.3	0.3	0.2	2.23	2	4	0.2	0.2	0.02	0.02	0.02	0.02
73	1.2	3.2	0.16	0.24	0.36	0.24	1.77	2	2	0.2	0.2	0.01	0.01	0.01	0.01
74	1.2	3.4	0.12	0.18	0.42	0.28	1.38	2	4	0.2	0.2	0.01	0.01	0.02	0.01
75	1.2	3.6	0.08	0.12	0.48	0.32	1.06	2	2	0.2	0.2	0.0	0.0	0.01	0.01
76	1.2	3.8	0.04	0.06	0.54	0.36	0.81	2	4	0.2	0.2	0.0	0.0	0.02	0.01
77	1.2	4.0	0.0	0.0	0.6	0.4	0.60	2	1	0.2	0.2	0.0	0.0	0.0	0.0
78	1.4	2.0	0.3	0.7	0.0	0.0	6.99	4	1	0.2	0.2	0.04	0.09	0.0	0.0
79	1.4	2.2	0.27	0.63	0.07	0.03	5.97	4	4	0.2	0.2	0.11	0.27	0.03	0.01
80	1.4	2.4	0.24	0.56	0.14	0.06	5.02	4	2	0.2	0.2	0.04	0.10	0.03	0.01
81	1.4	2.6	0.21	0.49	0.21	0.09	4.16	4	4	0.2	0.2	0.06	0.15	0.06	0.03
82	1.4	2.8	0.18	0.42	0.28	0.12	3.40	4	2	0.2	0.2	0.02	0.05	0.03	0.01
83	1.4	3.0	0.15	0.35	0.35	0.15	2.74	4	4	0.2	0.2	0.03	0.07	0.07	0.03
84	1.4	3.2	0.12	0.28	0.42	0.18	2.17	4	2	0.2	0.2	0.01	0.02	0.03	0.01
85	1.4	3.4	0.09	0.21	0.49	0.21	1.69	4	4	0.2	0.2	0.01	0.03	0.06	0.03
86	1.4	3.6	0.06	0.14	0.56	0.24	1.30	4	2	0.2	0.2	0.0	0.01	0.03	0.01
87	1.4	3.8	0.03	0.07	0.63	0.27	0.99	4	4	0.2	0.2	0.0	0.0	0.04	0.02
88	1.4	4.0	0.0	0.0	0.7	0.3	0.74	4	1	0.2	0.2	0.0	0.0	0.01	0.0
89	1.6	2.0	0.2	0.8	0.0	0.0	8.43	2	1	0.2	0.2	0.01	0.06	0.0	0.0
90	1.6	2.2	0.18	0.72	0.08	0.02	7.20	2	4	0.2	0.2	0.05	0.18	0.02	0.01
91	1.6	2.4	0.16	0.64	0.16	0.04	6.06	2	2	0.2	0.2	0.02	0.07	0.02	0.0
92	1.6	2.6	0.14	0.56	0.24	0.06	5.02	2	4	0.2	0.2	0.03	0.10	0.04	0.01
93	1.6	2.8	0.12	0.48	0.32	0.08	4.10	2	2	0.2	0.2	0.01	0.04	0.02	0.01

(Continued)

TABLE 5.22 (*Continued*)

Calculation of Arc Heat for Element 2 for t = 2 s

S. No.	x	y	N1	N2	N3	N4	q	Wx	Wy	hx	hy	F1	F2	F3	F4
94	1.6	3.0	0.1	0.4	0.4	0.1	3.30	2	4	0.2	0.2	0.01	0.05	0.05	0.01
95	1.6	3.2	0.08	0.32	0.48	0.12	2.62	2	2	0.2	0.2	0.0	0.01	0.02	0.01
96	1.6	3.4	0.06	0.24	0.56	0.14	2.04	2	4	0.2	0.2	0.0	0.02	0.04	0.01
97	1.6	3.6	0.04	0.16	0.64	0.16	1.57	2	2	0.2	0.2	0.0	0.0	0.02	0.0
98	1.6	3.8	0.02	0.08	0.72	0.18	1.19	2	4	0.2	0.2	0.0	0.0	0.03	0.01
99	1.6	4.0	0.0	0.0	0.8	0.2	0.89	2	1	0.2	0.2	0.0	0.0	0.01	0.0
100	1.8	2.0	0.1	0.9	0.0	0.0	10.01	4	1	0.2	0.2	0.02	0.16	0.0	0.0
101	1.8	2.2	0.09	0.81	0.09	0.01	8.56	4	4	0.2	0.2	0.05	0.49	0.05	0.01
102	1.8	2.4	0.08	0.72	0.18	0.02	7.20	4	2	0.2	0.2	0.02	0.18	0.05	0.01
103	1.8	2.6	0.07	0.63	0.27	0.03	5.97	4	4	0.2	0.2	0.03	0.27	0.11	0.01
104	1.8	2.8	0.06	0.54	0.36	0.04	4.87	4	2	0.2	0.2	0.01	0.09	0.06	0.01
105	1.8	3.0	0.05	0.45	0.45	0.05	3.92	4	4	0.2	0.2	0.01	0.13	0.13	0.01
106	1.8	3.2	0.04	0.36	0.54	0.06	3.11	4	2	0.2	0.2	0.0	0.04	0.06	0.01
107	1.8	3.4	0.03	0.27	0.63	0.07	2.43	4	4	0.2	0.2	0.01	0.05	0.11	0.01
108	1.8	3.6	0.02	0.18	0.72	0.08	1.87	4	2	0.2	0.2	0.0	0.01	0.05	0.01
109	1.8	3.8	0.01	0.09	0.81	0.09	1.41	4	4	0.2	0.2	0.0	0.01	0.08	0.01
110	1.8	4.0	0.0	0.0	0.9	0.1	1.06	4	1	0.2	0.2	0.0	0.0	0.02	0.0
111	2.0	2.0	0.0	1.0	0.0	0.0	11.72	1	1	0.2	0.2	0.0	0.05	0.0	0.0
112	2.0	2.2	0.0	0.9	0.1	0.0	10.01	1	4	0.2	0.2	0.0	0.16	0.02	0.0
113	2.0	2.4	0.0	0.8	0.2	0.0	8.43	1	2	0.2	0.2	0.0	0.06	0.01	0.0
114	2.0	2.6	0.0	0.7	0.3	0.0	6.99	1	4	0.2	0.2	0.0	0.09	0.04	0.0
115	2.0	2.8	0.0	0.6	0.4	0.0	5.71	1	2	0.2	0.2	0.0	0.03	0.02	0.0
116	2.0	3.0	0.0	0.5	0.5	0.0	4.59	1	4	0.2	0.2	0.0	0.04	0.04	0.0
117	2.0	3.2	0.0	0.4	0.6	0.0	3.64	1	2	0.2	0.2	0.0	0.01	0.02	0.0
118	2.0	3.4	0.0	0.3	0.7	0.0	2.84	1	4	0.2	0.2	0.0	0.02	0.04	0.0
119	2.0	3.6	0.0	0.2	0.8	0.0	2.18	1	2	0.2	0.2	0.0	0.0	0.02	0.0
120	2.0	3.8	0.0	0.1	0.9	0.0	1.66	1	4	0.2	0.2	0.0	0.0	0.03	0.0
121	2.0	4.0	0.0	0.0	1.0	0.0	1.24	1	1	0.2	0.2	0.0	0.0	0.01	0.0
Total												2.08	4.16	2.08	1.05

Note: x varies from 0 to 2 mm and y varies from 2 to 4 mm.

TABLE 5.23

Arc Heat Values Calculated for Various Time Steps for Element 2

Time, s	Heat Input, watts			
	Node 1	Node 2	Node 3	Node 4
0	0.32	0.11	0.05	0.16
0.125	0.69	0.26	0.13	0.35
0.25	1.38	0.56	0.28	0.69
0.375	2.50	1.13	0.57	1.25
0.5	4.16	2.08	1.05	2.08
0.625	6.33	3.54	1.78	3.17
0.75	8.82	5.52	2.77	4.42
0.875	11.26	7.91	3.97	5.65
1.0	13.17	10.40	5.21	6.61
1.125	14.13	12.55	6.30	7.09
1.25	13.90	13.90	6.97	6.97
1.375	12.55	14.13	7.09	6.30
1.5	10.40	13.17	6.61	5.21
1.625	7.91	11.26	5.65	3.97
1.75	5.52	8.82	4.42	2.77
1.875	3.54	6.33	3.17	1.78
2.0	2.08	4.16	2.08	1.05
2.125	1.13	2.50	1.25	0.57
2.25	0.56	1.38	0.69	0.28
2.375	0.26	0.69	0.35	0.13
2.5	0.11	0.32	0.16	0.05

The heat liberated at the nodes of the two elements is presented in Figure 5.30.

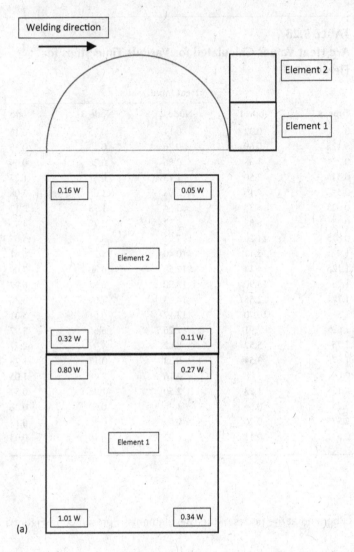

FIGURE 5.30 (a) The arc heat values at the four nodes of the element at time t = 0 s.

(*Continued*)

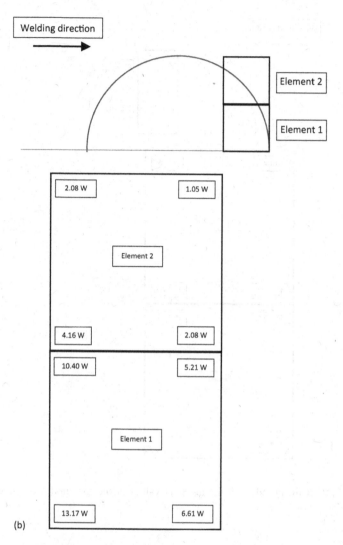

(b)

FIGURE 5.30 (Continued) (b) the arc heat values at the four nodes of the element at time t = 0.5 s.

(*Continued*)

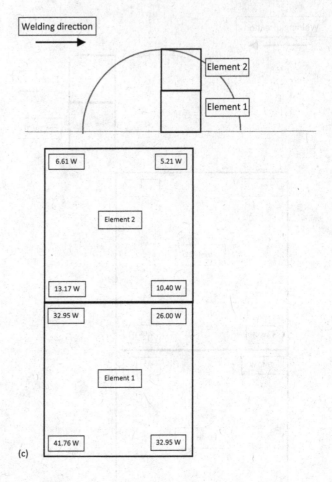

FIGURE 5.30 (Continued)   (c) the arc heat values at the four nodes of the element at
time t = 1 s.   (*Continued*)

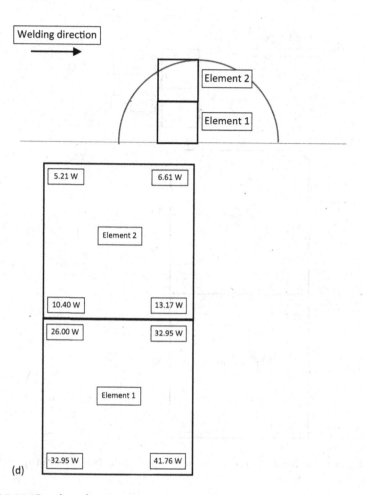

FIGURE 5.30 (Continued) (d) the arc heat values at the four nodes of the element at time t = 1.5 s.

(Continued)

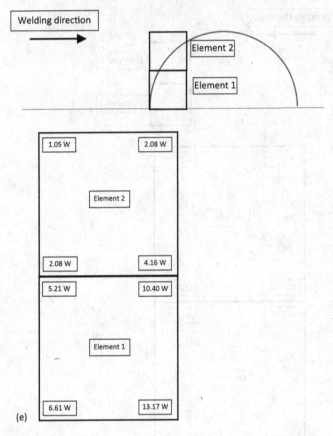

(e)

FIGURE 5.30 (Continued) (e) the arc heat values at the four nodes of the element at time t = 2 s.   *(Continued)*

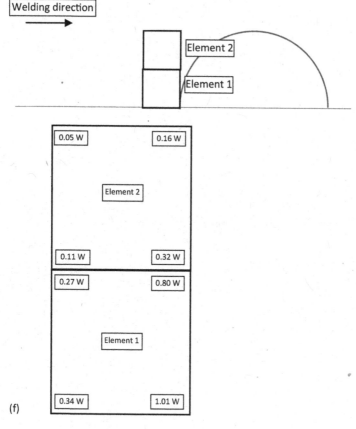

FIGURE 5.30 (Continued) (f) the arc heat values at the four nodes of the element at time t = 2.5 s.

REFERENCES

1. Friedman E. 1975. Thermomechanical analysis of welding process using finite element method. *Transactions of ASME Journal of Pressure Vessel Technology*, Vol. 97, pp. 206–213.
2. Krutz G W and Segerlind L J. 1978. Finite element analysis of welded structures. *Welding Journal*, Vol 57 (7), pp. 211s–216s.

6 Sample Problems

In the previous section, the calculation of arc heat for different time steps was explained in detail. This section explains the procedure for running the analysis for a welded plate. The list of programs using APDL language and some typical results are also presented.

6.1 CROSS SECTIONAL ANALYSIS OF A SUBMERGED ARC WELDED PLATE

Problem

Perform the cross sectional analysis of a submerged arc welded plate of dimensions $500 \times 300 \times 10$ mm. The welding is performed with the following parameters. The current is 300 A, voltage is 32 V, and welding speed is 5 mm/s. The welding process efficiency may be taken as 1.0. The thermal conductivity is 0.08 W/mm K, density is 8×10^{-6} kg/mm³, and specific heat is 600 J/kg K. The arc has a diameter of 10 mm. The arc is assumed to move by 1 mm for each time step.

Solution

To perform the analysis, the discretization of the cross section is performed in ANSYS either by direct generation of nodes or by the automatic meshing. Since the symmetry condition is applied, one half of the plate is taken for analysis. The mesh size in the weld region has been selected as 1 mm so that there are six nodes which are connected to five elements which receive the arc heat input. The element size is progressively increased from the weld region.

In the direct generation route, the nodes are generated using "n" command and multiple sets of nodes are generated using the "ngen" command. Similarly the element is defined using "e" command and multiple sets of elements are generated using the "egen" command. In the other route of meshing, the region is divided into many areas which are defined using keypoints and lines. Then the mesh spacing at various keypoints is defined using the "kesize" command. When the mesh is generated, the size requirement of elements is met near the various keypoints. In the present case, the mesh size of elements which come directly under the arc has been kept as 1 mm so that the six nodes located at the top surface receive the arc heat.

Subsequently, the arc heat is calculated for various time intervals. The arc is assumed to move by a distance of 1 mm during each time step, and the corresponding time increment is 0.2 s. At time $t = 0$, the nodes receive much less heat. The heat

TABLE 6.1

Calculated Arc Heat Input at the Cross Section for Various Time Steps

Time, s	Heat Input, watts					
	Node 1	Node 2	Node 3	Node 4	Node 5	Node 6
0	8.95	15.95	11.28	6.34	2.83	0.69
0.2	26.35	46.97	33.23	18.66	8.32	2.03
0.4	61.05	108.8	76.97	43.23	19.28	4.7
0.6	111.24	198.24	140.25	78.78	35.13	8.56
0.8	159.44	284.14	201.03	112.92	50.35	12.26
1.0	179.77	320.37	226.66	127.31	56.77	13.83
1.2	159.44	284.14	201.03	112.92	50.35	12.26
1.4	111.24	198.24	140.25	78.78	35.13	8.56
1.6	61.05	108.8	76.97	43.23	19.28	4.7
1.8	26.35	46.97	33.23	18.66	8.32	2.03
2.0	8.95	15.95	11.28	6.34	2.83	0.69

increases with time, and the maximum heat is experienced at time t = 1.0 s. At this time, the centre of the arc is directly over the section. Subsequently, the arc heat starts decreasing, and at time t = 2.0 s, the arc heat again is a low value. After 2.0 s, the section does not receive any heat input.

The calculation procedure of the heat input values for various time intervals has been explained in Section 5 (example problem 5.1.1). The calculated values are presented in Table 6.1.

In the above calculation, the time t has been used for locating the arc in the element. The arc is assumed to move by 1 mm for each time step. So at a particular time step, the arc moves to the new position and the heat input commences at the nodes. The time in the above calculation represents the period *before* the heat input is experienced at the nodes.

The time as defined in ANSYS is different from what has been used in the above calculation. The time step in ANSYS represents the period *after* the heat input is received at the nodes. The time step in ANSYS cannot begin with time t = 0 s, and the first time step has to be 0.2 s. Hence, the heat input from the calculation corresponding to 0 s must be assigned for a time step of 0.2 s in ANSYS and so on. Thus, there will be effectively an offset of one time increment between the calculation and ANSYS.

The arc heat is applied at the identified nodes using "f" command. The "if" command is used to apply the heat input values corresponding to various time intervals. The convection boundary condition is applied at the nodes which lie on the surfaces. The convection coefficient and ambient temperature are defined at each of the surface nodes using the "sf" command. The "solve" command is used for the determination of nodal temperatures for a time step and for solving multiple time steps, the "do" command is used. For each loop, the time is set using the "time" command. The time increment is 0.2 s initially during the arc crossing phase, and thereafter, the time is progressively increased. The analysis is run up to a time of 3420 s.

The APDL programs for the two cases are presented below. The temperature plot corresponding to various time intervals are presented in subsequent figures. Figures 6.1 through 6.7 show the results for the first case while Figures 6.8 through 6.14 shows the results for the second case.

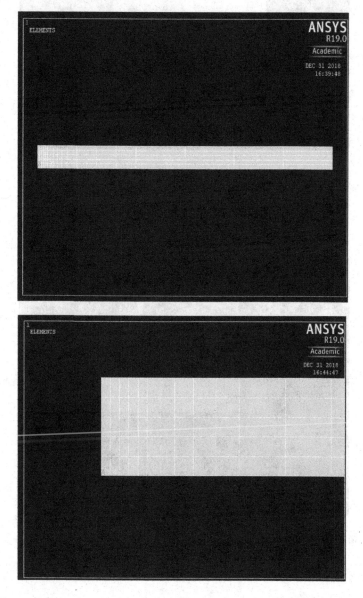

FIGURE 6.1 Discretization of the plate for cross sectional analysis.

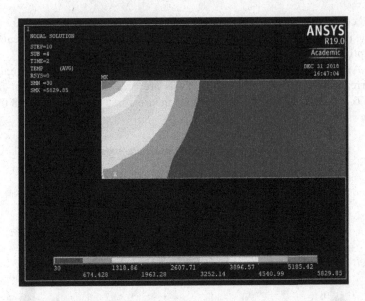

FIGURE 6.2 Temperature distribution in the plate at 2 s.

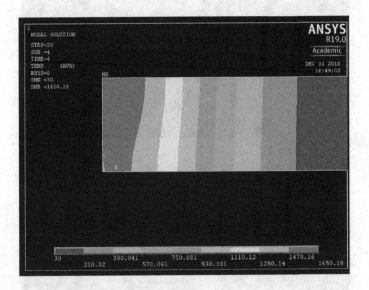

FIGURE 6.3 Temperature distribution in the plate at 4 s.

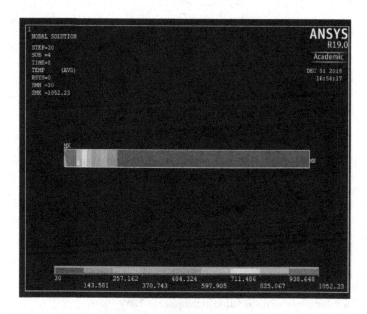

FIGURE 6.4 Temperature distribution in the plate at 8 s.

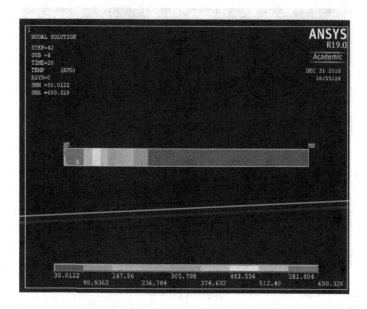

FIGURE 6.5 Temperature distribution in the plate at 20 s.

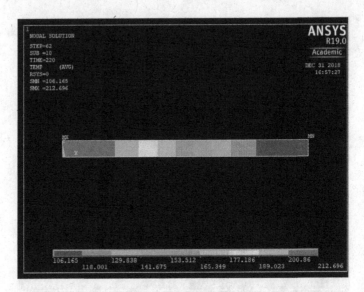

FIGURE 6.6   Temperature distribution in the plate at 220 s.

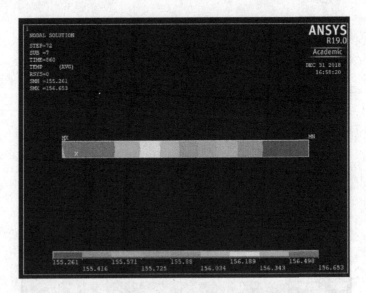

FIGURE 6.7   Temperature distribution in the plate at 860 s.

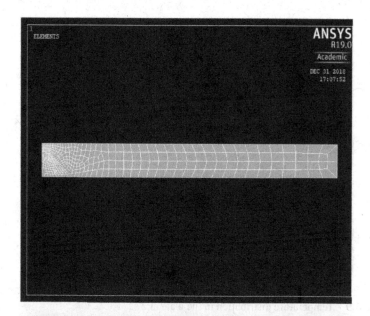

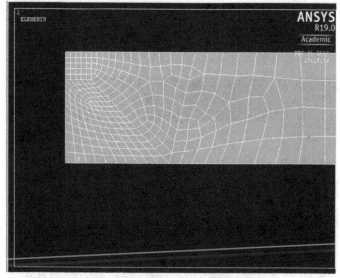

FIGURE 6.8 Discretization of the plate for cross sectional analysis.

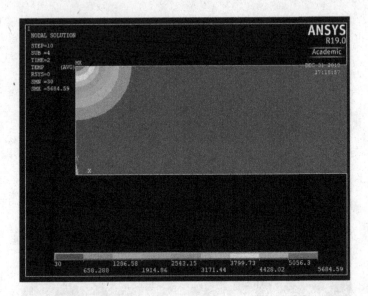

FIGURE 6.9   Temperature distribution in the plate at 2 s.

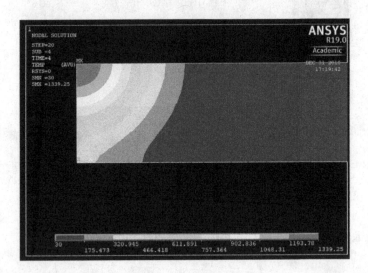

FIGURE 6.10   Temperature distribution in the plate at 4 s.

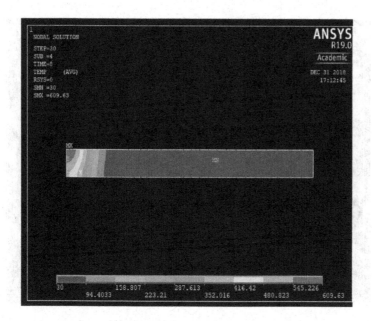

FIGURE 6.11   Temperature distribution in the plate at 8 s.

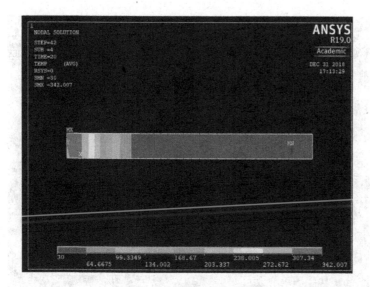

FIGURE 6.12   Temperature distribution in the plate at 20 s.

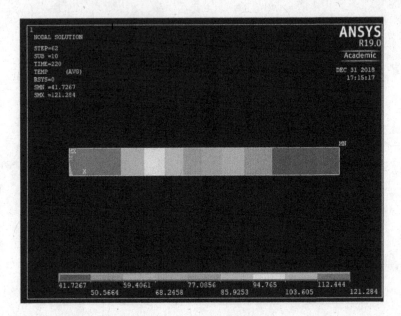

FIGURE 6.13   Temperature distribution in the plate at 220 s.

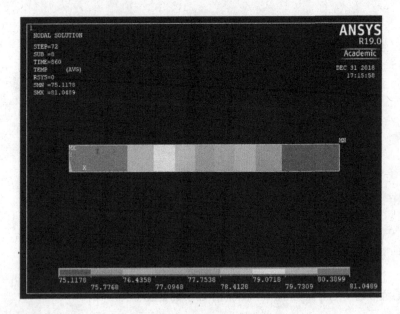

FIGURE 6.14   Temperature distribution in the plate at 860 s.

Case 1. Direct Generation of nodes.

```
/prep7

! defining the element type
et,1,plane55

! defining the material properties
! kxx is the thermal conductivity in W/mm K
! c is the specific heat in J / kg K
! dens is the density of the material in kg/mm3
mp,kxx,1,0.08
mp,c,1,600
mp,dens,1,8e-6

! defining the first set of nodes
n,1,0,10
n,6
fill

! generating multiple copies of nodes from the master set of nodes using the
ngen command
ngen,6,6,1,6,1,1
ngen,3,6,31,36,1,1.5
ngen,2,6,43,48,1,2
ngen,3,6,49,54,1,2.5
ngen,4,6,61,66,1,5
ngen,3,6,79,84,1,10
ngen,5,6,91,96,1,25

! defining the first element
e,2,8,7,1

! generating multiple copies of elements from the master set of elements
using the egen command
egen,5,1,-1
egen,19,6,-5
fini
/solu

! defining the analysis type as transient
antype,4
nropt,full
autots,on

! defining uniform temperature and reference temperature
tunif,30
tref,30
```

```
! defining the number of substeps in each time increment and the number
of iterations
nsubst,20
neqit,20
timint,on,ther
tintp,,,,1
eqslv,jcg

! defining convection boundary condition on nodes lying on the surface
nsel,y,0
sf,all,conv,1e-6,30
nall
nsel,x,150
sf,all,conv,1e-6,30
nall
nsel,node,31,145,6
sf,all,conv,1e-6,30
nall
fdele,all,all

! entering in the do loop for 11 time steps
! the arc heat is experienced only during these time steps
*do,i,1,11

! defining time step
time,i*0.2
fdele,all,all

! defining the heat input at the appropriate nodes
*if,i,eq,1,or,i,eq,11,then
f,1,heat,8.95
f,7,heat,15.95
f,13,heat,11.28
f,19,heat,6.34
f,25,heat,2.83
f,31,heat,0.69
*endif
*if,i,eq,2,or,i,eq,10,then
f,1,heat,26.35
f,7,heat,46.97
f,13,heat,33.23
f,19,heat,18.66
f,25,heat,8.32
f,31,heat,2.03
*endif
*if,i,eq,3,or,i,eq,9,then
f,1,heat,61.05
```

```
f,7,heat,108.8
f,13,heat,76.97
f,19,heat,43.23
f,25,heat,19.28
f,31,heat,4.7
*endif
*if,i,eq,4,or,i,eq,8,then
f,1,heat,111.24
f,7,heat,198.24
f,13,heat,140.25
f,19,heat,78.78
f,25,heat,35.13
f,31,heat,8.56
*endif
*if,i,eq,5,or,i,eq,7,then
f,1,heat,159.44
f,7,heat,284.14
f,13,heat,201.03
f,19,heat,112.92
f,25,heat,50.35
f,31,heat,12.26
*endif
*if,i,eq,6,then
f,1,heat,179.77
f,7,heat,320.37
f,13,heat,226.66
f,19,heat,127.31
f,25,heat,56.77
f,31,heat,13.83
*endif
solve
*enddo


! entering in the subsequent do loops
! there is no arc heat input in these time steps

*do,i,1,9
time,2.2+i*0.2
fdele,all,all
solve,
*enddo
*do,i,1,10
time,4+i*0.4
fdele,all,all
solve
*enddo
```

```
*do,i,1,12
time,8+i
fdele,all,all
solve
*enddo
*do,i,1,10
time,20+i*4
fdele,all,all
solve
*enddo
*do,i,1,10
time,60+i*16
fdele,all,all
solve
*enddo
*do,i,1,10
time,220+i*64
fdele,all,all
solve
*enddo
*do,i,1,10
time,860+i*256
fdele,all,all
solve
*enddo
fini
```

Case 2. Generation of meshing using solid modeling

```
/prep7

! defining the element type
et,1,plane55

! defining the material properties
! kxx is the thermal conductivity in W/mm K
! c is the specific heat in J/kg K
! dens is the density in kg/mm3
mp,kxx,1,0.08
mp,c,1,600
mp,dens,1,8e-6
```

```
! defining the keypoints
k,1,,20
k,2,,15
k,3
k,4,5,20
k,5,5,15
k,6,20
k,7,25,20
k,8,50,20
k,9,50
k,10,200,20
k,11,200

! defining the lines
l,4,7
l,7,8
l,8,10
l,10,11
l,9,11
l,6,9
l,3,6

! defining the areas
a,2,5,4,1
a,3,6,5,2
a,5,6,7,4
a,6,9,8,7
a,9,11,10,8

! specifying the desired element size near the keypoints
kesize,all,1
kesize,3,2
kesize,6,2
kesize,7,2.5
kesize,8,5
kesize,9,5
kesize,10,10
kesize,11,10

! meshing of the areas
amesh,all
fini
/solu
```

```
! defining analysis type as transient
antype,4

! defining uniform temperature and reference temperature
tunif,30
tref,30

! defining the number of substeps in each time increment and the number
of iterations
nsubst,10
neqit,10

! defining convection boundary condition on all surface nodes
*do,i,1,7
sfl,i,conv,1e-6,1e-6,30,30
*enddo
fdele,all,all

! entering in the do loop for the first 11 steps
! the arc heat is experienced only during these time steps
*do,i,1,11
time,i*0.2
fdele,all,all

! defining the heat input at the appropriate nodes
*if,i,eq,1,or,i,eq,11,then
f,12,heat,8.95
f,16,heat,15.95
f,15,heat,11.28
f,14,heat,6.34
f,13,heat,2.83
f,7,heat,0.69
*endif
*if,i,eq,2,or,i,eq,10,then
f,12,heat,26.35
f,16,heat,46.97
f,15,heat,33.23
f,14,heat,18.66
f,13,heat,8.32
f,7,heat,2.03
*endif
*if,i,eq,3,or,i,eq,9,then
f,12,heat,61.05
f,16,heat,108.8
f,15,heat,76.97
f,14,heat,43.23
```

```
f,13,heat,19.28
f,7,heat,4.7
*endif
*if,i,eq,4,or,i,eq,8,then
f,12,heat,111.24
f,16,heat,198.24
f,15,heat,140.25
f,14,heat,78.78
f,13,heat,35.13
f,7,heat,8.56
*endif
*if,i,eq,5,or,i,eq,7,then
f,12,heat,159.44
f,16,heat,284.14
f,15,heat,201.03
f,14,heat,112.92
f,13,heat,50.35
f,7,heat,12.26
*endif
*if,i,eq,6,then
f,12,heat,179.77
f,16,heat,320.37
f,15,heat,226.66
f,14,heat,127.31
f,13,heat,56.77
f,7,heat,13.83
*endif
solve
*enddo


! entering in the subsequent do loops
! there are no arc heat inputs during these time steps
*do,i,1,9
time,2.2+i*0.2
fdele,all,all
solve
*enddo
*do,i,1,10
time,4+i*0.4
fdele,all,all
solve
*enddo
*do,i,1,12
time,8+i
fdele,all,all
solve
```

```
*enddo
*do,i,1,10
time,20+i*4
fdele,all,all
solve
*enddo
*do,i,1,10
time,60+i*16
fdele,all,all
solve
*enddo
*do,i,1,10
time,220+i*64
fdele,all,all
solve
*enddo
*do,i,1,10
time,860+i*256
fdele,all,all
solve
*enddo
fini
```

6.2 IN-PLANE ANALYSIS OF A GAS METAL ARC WELDED PLATE

Problem

Perform the in-plane thermal analysis for the following case.

Welding process = Gas metal arc welding
Welding current = 100 A
Welding voltage = 20 V
Process efficiency = 0.75
Welding speed = 5 mm/s
Arc diameter = 10 mm
Plate size = 500 mm long, 200 mm wide, and 3 mm thick
Element width in the weld zone = 5 × 5 mm
Distance moved by the arc in one time step = 1 mm

Since the plate thickness is only 3 mm, a two-dimensional in-plane type of analysis is the appropriate choice for the problem. The symmetry condition has been invoked and hence, only a half section is taken for the analysis.

The material properties thermal conductivity and specific heat are assumed to be temperature dependent as given in Table 6.2.

The next step is to calculate the heat input values which are experienced by an element which lies along the weld length. The time taken by the arc to cross an element is 3 s. The time increment during the welding phase is determined from the distance

TABLE 6.2
Temperature Dependent Material Properties Used in the Analysis

Temperature, °C	Thermal Conductivity, W/mm K	Specific Heat, J/kg K	Density, kg/mm³
0	0.078	200	8×10^{-6}
500	0.058	600	8×10^{-6}
1200	0.03	600	8×10^{-6}
1538	0.03	600	8×10^{-6}
1540	0.06	600	8×10^{-6}
3000	0.06	600	8×10^{-6}

TABLE 6.3
Heat Input Values for Various Time Steps

Time, s	Heat Input, watts			
	Node 1	Node 2	Node 3	Node 4
0	3.15	0.48	0.22	1.44
0.2	10.73	1.95	0.89	4.91
0.4	29.42	6.47	2.96	13.47
0.6	65.05	17.75	8.13	29.78
0.8	116.76	40.67	18.62	53.45
1.0	171.44	78.49	35.93	78.49
1.2	207.78	128.38	58.77	95.12
1.4	209.89	178.36	81.65	96.08
1.6	178.36	209.89	96.08	81.65
1.8	128.38	207.78	95.12	58.77
2.0	78.49	171.44	78.49	35.93
2.2	40.67	116.76	53.45	18.62
2.4	17.75	65.05	29.78	8.13
2.6	6.47	29.42	13.47	2.96
2.8	1.95	10.73	4.91	0.89
3.0	0.48	3.15	1.44	0.22

moved by the arc for a given time step. In the present case, the time increment is
0.2 s. Thus the heat input values have been calculated for the duration of 3 s with a
time increment of 0.2 s as explained in Section 5 (example problem 5.2.2). The values which are calculated are given in Table 6.3.

The time in the above calculation is used for fixing the position of the arc and for
determining the arc heat liberated in the element. The position of the arc in the plate
for various time steps is as shown in Figure 6.15. The corresponding heat input at the
elements is also indicated in the figures.

In a set of elements which lie along the welding direction as shown, the arc is distributed over two elements, elements 1 and 2 at t = 0 s, and the arc is about to enter in element 3. From the position of the arc, it can be seen that the corresponding time for elements 3, 2 and 1 are 0, 1 and 2 s, respectively. After 1 s, the arc completely moves over elements 2 and 3.

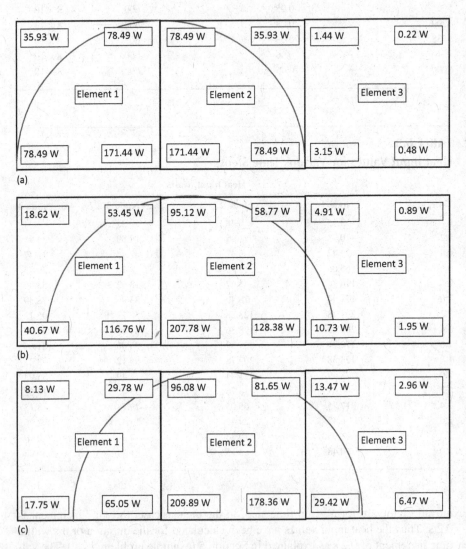

FIGURE 6.15 (a) Arc heat distribution in the elements corresponding to t = 0 s, (b) arc heat distribution in the elements corresponding to t = 0.2 s, (c) arc heat distribution in the elements corresponding to t = 0.4 s.   *(Continued)*

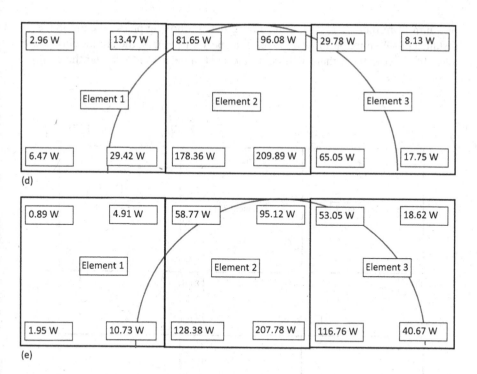

(d)

(e)

FIGURE 6.15 (Continued) (d) arc heat distribution in the elements corresponding to t = 0.6 s, and (e) arc heat distribution in the elements corresponding to t = 0.8 s.

Figure 6.16. shows the total nodal heat input for various positions of the arc. The sum of the heat input at various nodes is slightly less than the theoretical value as the stray heat flux which falls outside the arc diameter is not accounted in the calculation.

The pattern gets repeated after 1 s and the arc heat values shift to the next set of nodes. Thus if the node numbers are known, then it is possible to determine the heat input especially when the analysis is carried out with the help of a program.

The discretization of the plate into various elements and nodes is performed using direct generation of nodes and elements as shown in Figure 6.17. The length of the plate, which is along the x direction, is divided into 100 equal segments of 5 mm each. In the y direction, the element size has been selected to be 5 mm near the weld. The spacing is gradually increased, and there are 10 nodes in the y direction. Thus, there are a total of 1010 nodes and 900 elements in the plate.

Initially, the arc has its centre at a distance of 5 mm from the edge as shown in Figure 6.18. Similarly, during the last step, the arc has its centre at a distance of 5 mm from the other edge. Thus during the arc moving phase, the arc moves a distance of 490 mm. As the arc moves at a speed of 5 mm/s, the arc moving phase takes 98 s. After 98 s, there will be no arc heat input in the plate and the cooling phase begins.

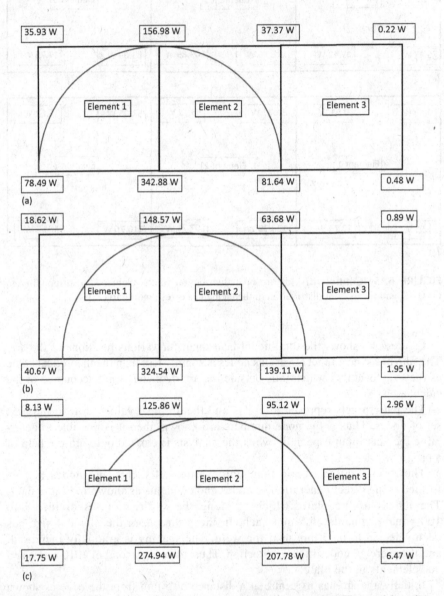

FIGURE 6.16 (a) Nodal heat input corresponding to time t = 0 s. The total heat input at various nodes adds to 733.99 W, (b) nodal heat input corresponding to time t = 0.2 s. The total heat input at various nodes adds to 738.03 W, (c) nodal heat input corresponding to time t = 0.4 s. The total heat input at various nodes adds to 739.01 W.   (*Continued*)

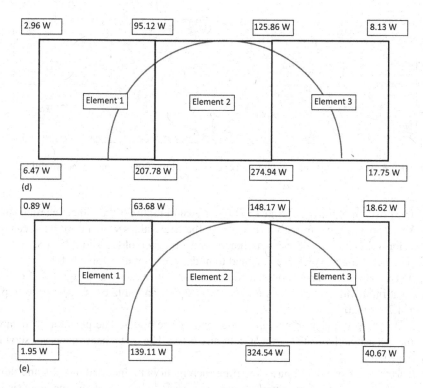

FIGURE 6.16 (Continued) (d) nodal heat input corresponding to time t = 0.6 s. The total heat input at various nodes adds to 739.01 W, and (e) nodal heat input corresponding to time t = 0.8 s. The total heat input at various nodes adds to 737.63 W.

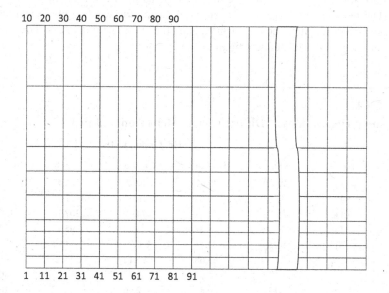

FIGURE 6.17 Discretization of the plate for analysis. Picture is not to scale.

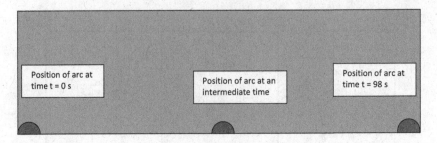

FIGURE 6.18   The starting, middle, and ending position of the arc along the weld line.

The position of the arc at time t = 0 s is shown in Figure 6.19a. Since the time step in ANSYS cannot begin with a time t = 0, the heat values which have been calculated for a time t = 0 have been assigned for a time step of 0.2 s in ANSYS, and so on. The heat values which are assigned to nodes are shown in Table 6.4.

At 1 s, the arc advances to the next set of elements as shown in Figure 6.19b and the heat input values at various nodes are as shown in Table 6.5 for the time steps from 1.2 s to 2.0 s.

In this way, the heat input values are input at the nodes. The position of the arc for the last set of elements is shown in Figure 6.19c. The nodal heat values are shown in Table 6.6.

During arc movement phase, the time increment is maintained at 0.2 s and during cooling phase, the time intervals are progressively increased and the analysis is performed up to a time of 5560 s.

The APDL program listing for the problem is presented below. The temperature plots corresponding to various time intervals are as shown in Figures 6.20 through 6.30. The peak temperature of the weld pool during the welding phase is 1990°C. At 98 s, the arc heat input comes to an end and the cooling commences. The cooling pattern is seen in the subsequent figures.

TABLE 6.4
The Nodal Heat Values at Different Time Steps from 0.2 to 1 s

		Heat Input, watts				
S. No.	Node	T = 0.2 s	T = 0.4 s	T = 0.6 s	T = 0.8 s	t = 1.0 s
1	1	78.49	40.67	17.75	6.47	1.95
2	2	35.93	18.62	8.13	2.96	0.89
3	11	342.88	324.54	274.94	207.78	139.11
4	12	156.98	148.57	125.86	95.12	63.68
5	21	81.64	139.11	207.78	274.94	324.54
6	22	37.37	63.68	95.12	125.86	148.57
7	31	0.48	1.95	6.47	17.75	40.67
8	32	0.22	0.89	2.96	8.13	18.62

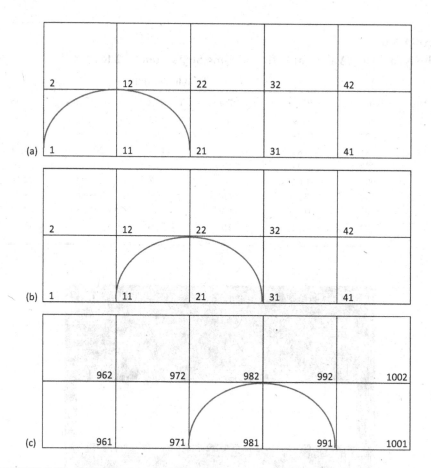

FIGURE 6.19   (a) The position of the arc at time t = 0 s, (b) the position of the arc at time
t = 1 s, and (c) the position of the arc at time t = 97 s.

TABLE 6.5
The Nodal Heat Values at Different Time Steps from 1.2 to 2 s

		Heat Input, watts				
S. No.	Node	T = 1.2 s	T = 1.4 s	T = 1.6 s	T = 1.8 s	t = 2.0 s
1	11	78.49	40.67	17.75	6.47	1.95
2	12	35.93	18.62	8.13	2.96	0.89
3	21	342.88	324.54	274.94	207.78	139.11
4	22	156.98	148.57	125.86	95.12	63.68
5	31	81.64	139.11	207.78	274.94	324.54
6	32	37.37	63.68	95.12	125.86	148.57
7	41	0.48	1.95	6.47	17.75	40.67
8	42	0.22	0.89	2.96	8.13	18.62

TABLE 6.6

The Nodal Heat Values at Different Time Steps from 97.2 to 98 s

S. No.	Node	T = 97.2 s	T = 97.4 s	T = 97.6 s	T = 97.8 s	t = 98 s
				Heat Input, watts		
1	971	78.49	40.67	17.75	6.47	1.95
2	972	35.93	18.62	8.13	2.96	0.89
3	981	342.88	324.54	274.94	207.78	139.11
4	982	156.98	148.57	125.86	95.12	63.68
5	991	81.64	139.11	207.78	274.94	324.54
6	992	37.37	63.68	95.12	125.86	148.57
7	1001	0.48	1.95	6.47	17.75	40.67
8	1002	0.22	0.89	2.96	8.13	18.62

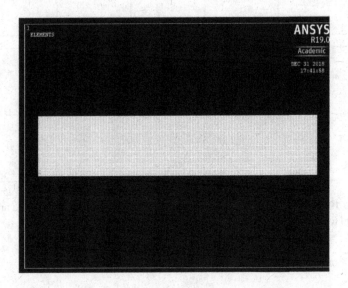

FIGURE 6.20   Discretization of the plate for analysis.

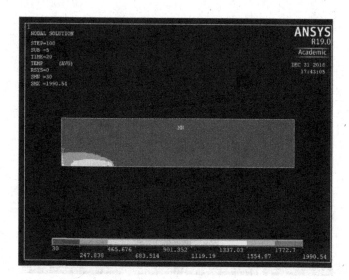

FIGURE 6.21 Temperature distribution in the plate at 20 s.

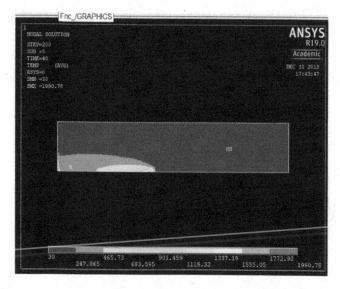

FIGURE 6.22 Temperature distribution in the plate at 40 s.

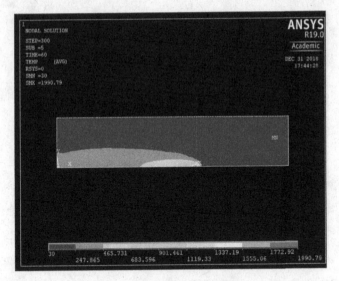

FIGURE 6.23   Temperature distribution in the plate at 60 s.

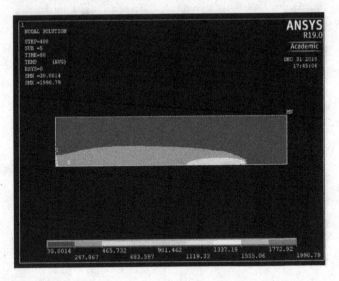

FIGURE 6.24   Temperature distribution in the plate at 80 s.

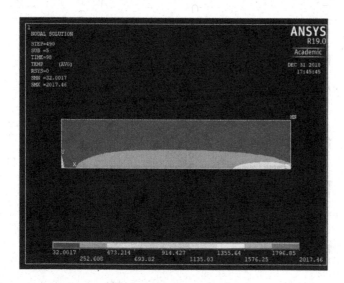

FIGURE 6.25   Temperature distribution in the plate at 98 s.

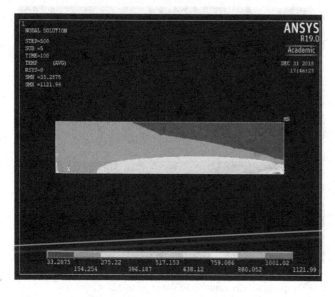

FIGURE 6.26   Temperature distribution in the plate at 100 s.

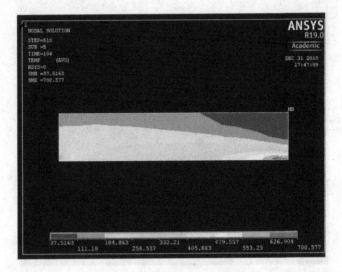

FIGURE 6.27 Temperature distribution in the plate at 104 s.

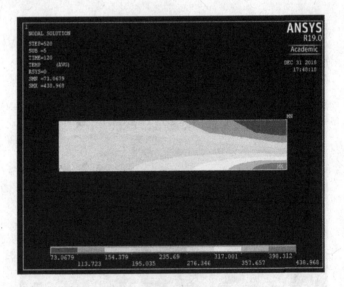

FIGURE 6.28 Temperature distribution in the plate at 120 s.

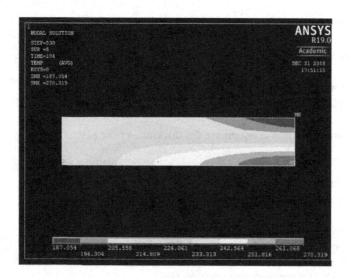

FIGURE 6.29 Temperature distribution in the plate at 184 s.

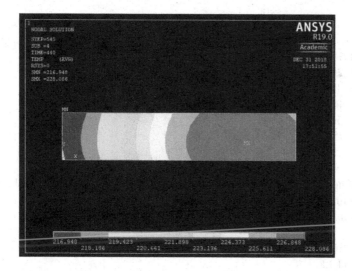

FIGURE 6.30 Temperature distribution in the plate at 440 s.

The APDL program for the case of MIG welded plate

```
/prep7

! defining the element type
et,1,plane55,,,3

! defining the real constant viz the thickness of the plate
r,1,3

! defining the material properties as a function of temperature
! kxx is the thermal conductivity in W/mm K
! c is the specific heat in J/kg K
! dens is the density in kg/mm3
mptemp,1,0,500,1200,1538,1540,3000
mpdata,kxx,1,1,0.078,0.058,0.03,0.03,0.06,0.06
mpdata,c,1,1,200,600,600,600,600,600
mpdata,dens,1,1,8e-6,8e-6,8e-6,8e-6,8e-6,8e-6

! defining the first set of nodes
n,1
n,2,,5
n,3,,10
n,4,,15
n,5,,20
n,6,,30
n,7,,40
n,8,,50
n,9,,75
n,10,,100

! generating multiple copies of nodes from the master set using the ngen
command
ngen,101,10,1,10,1,5

! defining the first element
e,1,11,12,2

! generating multiple copies of elements from the master set using the egen
command
egen,9,1,-1
egen,100,10,-9
fini
/solu
```

```
! defining the analysis type as transient
antype,4
nropt,full
autots,on

! defining uniform temperature and reference temperature
tunif,30
tref,30

! defining the number of substeps in each time increment and the number
of iterations
nsubst,20
neqit,20
timint,on,ther
tintp,,,,1
eqslv,jcg

! defining convection boundary condition on all nodes
sf,all,conv,5e-6,30
fdele,all,all

! entering in the do loop for first 490 steps
! the arc is moving along the plate during these time steps
*do,i,1,98

! defining the nodes and the corresponding arc heat input
n1=(i-1)*10+1
n2=i*10+1
n3=(i+1)*10+1
n4=(i+2)*10+1
n5=(i-1)*10+2
n6=i*10+2
n7=(i+1)*10+2
n8=(i+2)*10+2
*do,j,1,5

! defining the time increment
time,(i-1)+j*0.2
*if,j,eq,1,then
f1=78.49
f2=342.88
f3=81.64
f4=0.48
f5=35.93
```

```
f6=156.98
f7=37.37
f8=0.22
*endif
*if,j,eq,2,then
f1=40.67
f2=324.54
f3=139.11
f4=1.95
f5=18.62
f6=148.57
f7=63.68
f8=0.89
*endif
*if,j,eq,3,then
f1=17.75
f2=274.94
f3=207.78
f4=6.47
f5=8.13
f6=125.86
f7=95.12
f8=2.96
*endif
*if,j,eq,4,then
f1=6.47
f2=207.78
f3=274.94
f4=17.75
f5=2.96
f6=95.12
f7=125.86
f8=8.13
*endif
*if,j,eq,5,then
f1=1.95
f2=139.11
f3=324.54
f4=40.67
f5=0.89
f6=63.68
f7=148.57
f8=18.62
*endif
fdele,all,all
f,n1,heat,f1
```

```
f,n2,heat,f2
f,n3,heat,f3
f,n4,heat,f4
f,n5,heat,f5
f,n6,heat,f6
f,n7,heat,f7
f,n8,heat,f8
solve
*enddo
*enddo

! entering the subsequent do loops
! there is no arc heat during these subsequent time steps
*do,i,1,10
time,98+i*0.2
fdele,all,all
solve
*enddo
*do,i,1,10
time,100+i*0.4
fdele,all,all
solve
*enddo
*do,i,1,10
time,104+i*1.6
fdele,all,all
solve
*enddo
*do,i,1,10
time,120+i*6.4
fdele,all,all
solve
*enddo
*do,i,1,10
time,184+i*25.6
fdele,all,all
solve
*enddo
*do,i,1,10
time,440+i*102.4
*enddo
*do,i,1,10
time,1464+i*409.6
*enddo
fini
```

6.3 IN-PLANE ANALYSIS OF A GAS METAL ARC WELDED DISSIMILAR WELDMENT

Problem

Perform the analysis for the following case.

Welding process = Gas metal arc welding
Nature of welding = Dissimilar welding of carbon steel and stainless steel plates
Welding current = 150 A
Welding voltage = 23 V
Process efficiency = 0.7
Arc diameter = 10 mm
Welding speed = 5 mm/s
Plate size = 500 × 100 × 5 mm of carbon steel and 500 × 100 × 5 mm of stainless steel
Element size near the arc zone = 10 × 5 mm
Distance moved by the arc for one time step = 1 mm

Solution

Since different materials are welded, the symmetry condition cannot be invoked, and a full model has to be employed in this case. The plate thickness is only 5 mm, and the in-plane analysis is suitable for the problem.

The element generation has been performed in such a way that the elements lying on either side of the centre line of the plate are of carbon steel material and stainless steel material, respectively. The material properties of the two materials are taken to be temperature dependent and are presented in Table 6.7.

TABLE 6.7
Material Properties of the Materials at Various Temperatures

		Carbon Steel		Stainless Steel	
S. No.	Temperature °C	Thermal Conductivity, W/mm K	Specific Heat, J/kg K	Thermal Conductivity, W/mm K	Specific Heat, J/kg K
1	0	0.078	200	0.018	300
2	500	0.058	600	0.023	600
3	1200	0.03	600	0.03	600
4	1538	0.03	600	0.03	600
5	1540	0.06	600	0.06	600
6	3000	0.06	600	0.06	600

TABLE 6.8
Arc Heat Input Values for Different Time Steps

Time, s	Heat Input, watts			
	Node 1	Node 2	Node 3	Node 4
0	5.51	0.38	0.17	2.52
0.2	18.92	1.55	0.71	8.66
0.4	52.58	5.20	2.38	24.07
0.6	119.00	14.42	6.60	54.48
0.8	221.14	33.53	15.35	101.24
1.0	341.66	66.56	30.47	156.41
1.2	446.60	115.18	52.73	204.45
1.4	505.31	177.68	81.34	231.33
1.6	508.54	249.87	114.39	232.81
1.8	467.94	326.69	149.56	214.22
2.0	402.35	402.35	184.19	184.19
2.2	326.69	467.94	214.22	149.56
2.4	249.87	508.54	232.81	114.39
2.6	177.68	505.31	231.33	81.34
2.8	115.18	446.60	204.45	52.73
3.0	66.56	341.66	156.41	30.47
3.2	33.53	221.14	101.24	15.35
3.4	14.42	119.00	54.48	6.60
3.6	5.20	52.58	24.07	2.38
3.8	1.55	18.92	8.66	0.71
4.0	0.38	5.51	2.52	0.17

In the above case, the melting point has been assumed to be 1540°C for both the materials. The density of the materials has been assumed to be 8×10^{-6} kg/mm^3.

The heat input values for the element which lies along the weld line have been calculated as discussed in Section 5 (example problem 5.2.1) and are presented in Table 6.8.

Since this problem considers a full model, the arc heat spreads over four elements at any point of time. Initially, at t = 0, the arc heat lies entirely in elements 1 and 4 as shown. The heat liberated is symmetrical about the weld centreline. As the arc advances, elements 2 and 5 also experience the arc heat. At 2 s, the arc is entirely over elements 2 and 5. The position occupied by the arc at various time steps is shown in Figure 6.31.

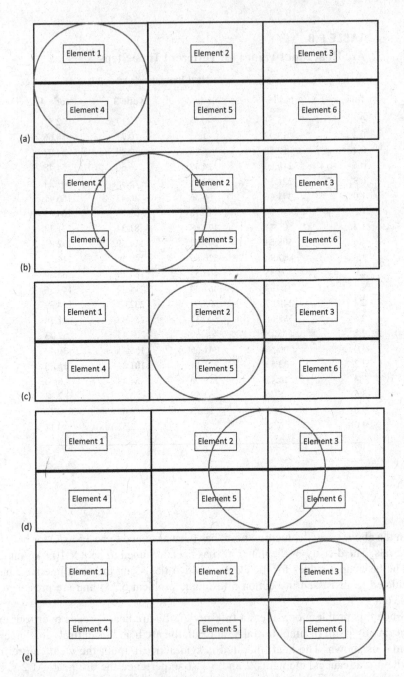

FIGURE 6.31 (a) Position occupied by the arc at time t = 0 s, (b) position occupied by the arc at time t = 1 s, (c) position occupied by the arc at time t = 2 s, (d) position occupied by the arc at time t = 3 s, and (e) position occupied by the arc at time t = 4 s.

From the position of the arc, it can be seen that the heat liberated in elements 2 and 1 for any position of the arc have a time lag of 2 s. Thus when the arc is about to enter in element 2, the heat input in element 2 corresponds to a time step of 0 s. The heat liberated in element 1 for that position of the arc corresponds to a time step of 2 s. The heat value in the elements 1 and 2 for various positions of the arc are taken from Table 6.8 and are as given in Figure 6.32.

The heat liberated at the nodes is summed up, and the nodal heat input is shown in Figure 6.33 for various positions of the arc.

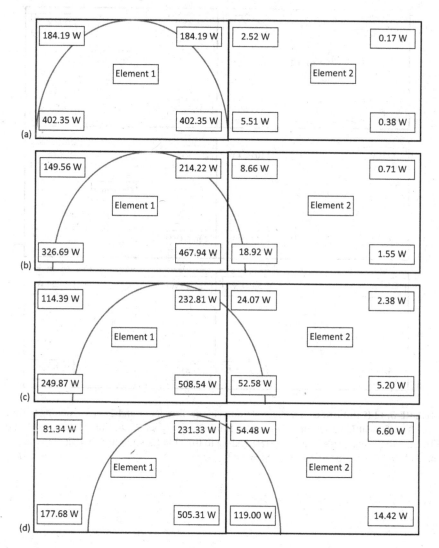

FIGURE 6.32 (a) Heat liberated in the elements 1 and 2 corresponding to time t = 0 s, (b) heat liberated in the elements 1 and 2 corresponding to time t = 0.2 s, (c) heat liberated in the elements 1 and 2 corresponding to time t = 0.4 s, (d) heat liberated in the elements 1 and 2 corresponding to time t = 0.6 s. *(Continued)*

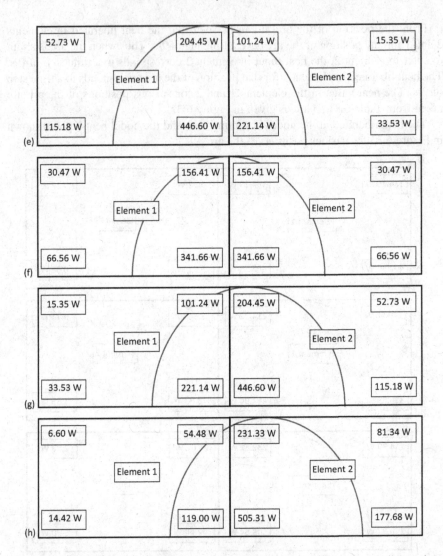

FIGURE 6.32 (Continued)   (e) heat liberated in the elements 1 and 2 corresponding to time t = 0.8 s, (f) heat liberated in the elements 1 and 2 corresponding to time t = 1.0 s, (g) heat liberated in the elements 1 and 2 corresponding to time t = 1.2 s, (h) heat liberated in the elements 1 and 2 corresponding to time t = 1.4 s.   (*Continued*)

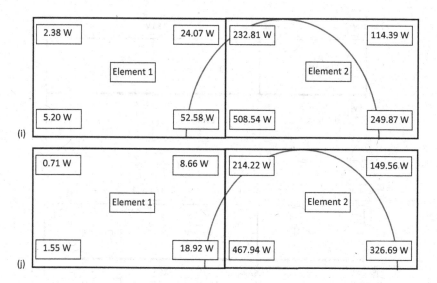

FIGURE 6.32 (Continued) (i) heat liberated in the elements 1 and 2 corresponding to time t = 1.6 s, and (j) heat liberated in the elements 1 and 2 corresponding to time t = 1.8 s.

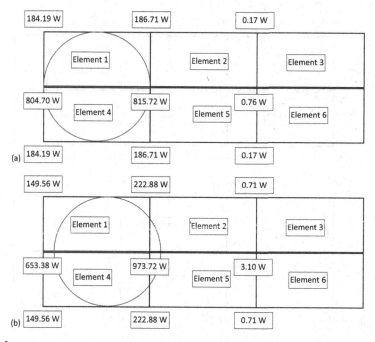

FIGURE 6.33 (a) Heat liberated at various nodes for time t = 0 s. The total heat input at various nodes is 2363.32 W, (b) heat liberated at various nodes for time t = 0.2 s. The total heat input at various nodes is 2376.5 W. *(Continued)*

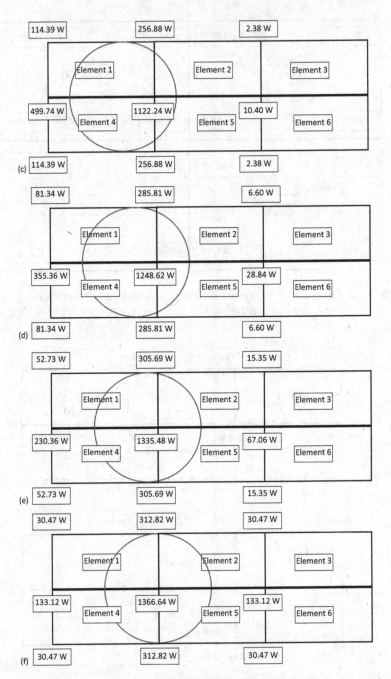

FIGURE 6.33 (Continued) (c) heat liberated at various nodes for time t = 0.4 s. The total heat input at various nodes is 2379.68 W. (d) heat liberated at various nodes for time t = 0.6 s. The total heat input at various nodes is 2380.32 W, (e) heat liberated at various nodes for time t = 0.8 s. The total heat input at various nodes is 2380.44 W, (f) heat liberated at various nodes for time t = 1.0 s. The total heat input at various nodes is 2380.4 W. *(Continued)*

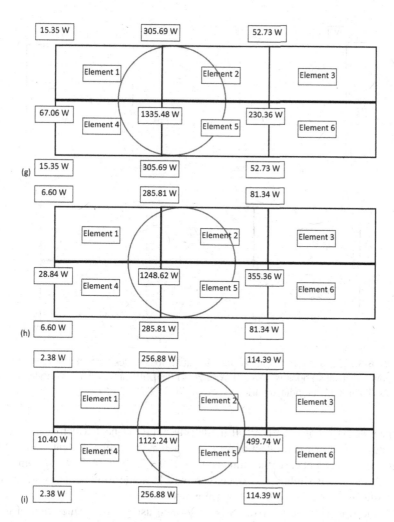

(g)

(h)

(i)

FIGURE 6.33 (Continued) (g) heat liberated at various nodes for time t = 1.2 s. The total heat input at various nodes is 2380.44 W, (h) heat liberated at various nodes for time t = 1.4 s. The total heat input at various nodes is 2380.32 W, (i) heat liberated at various nodes for time t = 1.6 s. The total heat input at various nodes is 2379.68 W. *(Continued)*

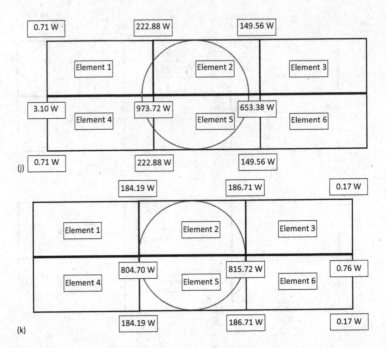

FIGURE 6.33 (Continued) (j) heat liberated at various nodes for time t = 1.8 s. The total heat input at various nodes is 2376.50 W, and (k) heat liberated at various nodes for time t = 2.0 s. The total heat input at various nodes is 2363.32 W.

Since we have considered a full model, the total arc heat which is liberated in the nodes should be the product of process efficiency, welding current, and welding voltage. In the present case, the value is 0.7 × 150 × 23 V = 2415 W. The sum of the calculated values is slightly less since there will be some stray heat flux which falls outside the arc diameter which is not accounted.

The mesh generation in ANSYS is performed using direct generation of nodes and elements as shown in Figure 6.34. In the length direction, the spacing has been maintained at 10 mm and there are 50 total elements for a length of 500 mm. In the width direction, the node spacing has been progressively increased from 5 mm, and there are 12 elements in the width direction. There are 663 nodes and 600 elements in total.

Initially, the arc has its centre at a distance of 5 mm from the edge as shown in Figure 6.35. Similarly, during the last step, the arc has its centre at a distance of 5 mm from the other edge. Thus during the arc moving phase, the arc moves a distance of 490 mm. As the arc moves at a speed of 5 mm/s, the arc moving phase will take 98 s. After 98 s, there will be no arc heat input in the plate and the cooling phase begins.

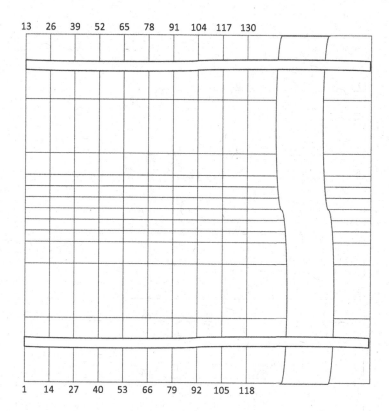

FIGURE 6.34 Discretization of the plate for analysis. Picture is not to scale.

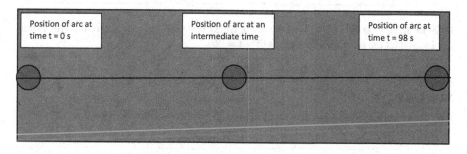

FIGURE 6.35 The starting, middle, and ending position of the arc along the weld line.

The position of the arc at time t = 0 s is shown in Figure 6.36a. Since the time step in ANSYS cannot begin with a time t = 0, the heat values which have been calculated for a time t = 0 have been assigned for a time step of 0.2 s and so on. The various heat values which are assigned to nodes are as shown in Table 6.9.

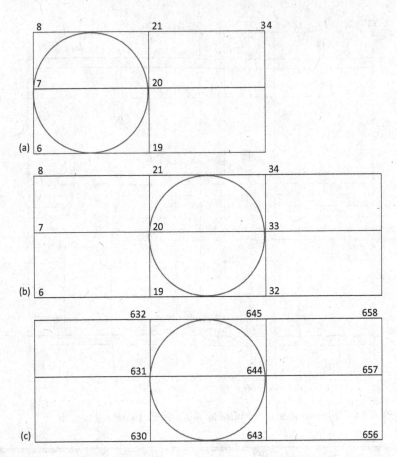

FIGURE 6.36 (a) The position of the arc at time t = 0 s, (b) the position of the arc at time t = 2 s, and (c) the position of the arc at time t = 96 s.

TABLE 6.9
Arc Heat Input at the Nodes for Various Time Steps from 0.2 to 2 s

		Heat Input, watts									
S. No.	Node	0.2 s	0.4 s	0.6 s	0.8 s	1.0 s	1.2 s	1.4 s	1.6 s	1.8 s	2.0 s
1	6	184.19	149.56	114.39	81.34	52.73	30.47	15.35	6.60	2.38	0.71
2	7	804.70	653.38	499.74	355.36	230.36	133.12	67.06	28.84	10.40	3.10
3	8	184.19	149.56	114.39	81.34	52.73	30.47	15.35	6.60	2.38	0.71
4	19	186.71	222.88	256.88	285.81	305.69	312.82	305.69	285.81	256.88	222.88
5	20	815.72	973.72	1122.24	1248.62	1335.48	1366.64	1335.48	1248.62	1122.24	973.72
6	21	186.71	222.88	256.88	285.81	305.69	312.82	305.69	285.81	256.88	222.88
7	32	0.17	0.71	2.38	6.60	15.35	30.47	52.73	81.34	114.39	149.56
8	33	0.76	3.10	10.40	28.84	67.06	133.12	230.36	355.36	499.74	653.38
9	34	0.17	0.71	2.38	6.60	15.35	30.47	52.73	81.34	114.39	149.56

At 2 s, the arc advances to the next set of elements as shown in Figure 6.36b and the heat input values at various nodes are as shown in Table 6.10 for the time steps from 2.2 s to 4.0 s.

The movement of the arc into the last set of elements corresponding to a time steps from 96.2 s to 98 s is as shown in Figure 6.36c. The corresponding arc heat input is shown in Table 6.11.

The APDL program for the given problem is presented below. The temperature results from ANSYS are presented in Figures 6.37 through 6.47. The peak temperature in the weld zone is 1743°C during the arc moving phase. At 98 s, the arc has reached the other edge of the plate, and welding phase comes to an end at this time step. For the subsequent time steps, there is no arc heat and the peak temperature gradually falls with time. The equalization of temperature is faster in the carbon steel side as the thermal conductivity is higher. The equalization takes a long time in the stainless steel side.

TABLE 6.10
Arc Heat Input at the Nodes for Various Time Steps from 2.2 to 4 s

		Heat Input, watts								
S. No.	Node	2.2 s	2.4 s	2.6 s	2.8 s	3.0 s	3.2 s	3.4 s	3.6 s	3.8 s	4.0 s
1	19	184.19	149.56	114.39	81.34	52.73	30.47	15.35	6.60	2.38	0.71
2	20	804.70	653.38	499.74	355.36	230.36	133.12	67.06	28.84	10.40	3.10
3	21	184.19	149.56	114.39	81.34	52.73	30.47	15.35	6.60	2.38	0.71
4	32	186.71	222.88	256.88	285.81	305.69	312.82	305.69	285.81	256.88	222.88
5	33	815.72	973.72	1122.24	1248.62	1335.48	1366.64	1335.48	1248.62	1122.24	973.72
6	34	186.71	222.88	256.88	285.81	305.69	312.82	305.69	285.81	256.88	222.88
7	45	0.17	0.71	2.38	6.60	15.35	30.47	52.73	81.34	114.39	149.56
8	46	0.76	3.10	10.40	28.84	67.06	133.12	230.36	355.36	499.74	653.38
9	47	0.17	0.71	2.38	6.60	15.35	30.47	52.73	81.34	114.39	149.56

TABLE 6.11
Arc Heat Input at the Nodes for Various Time Steps from 96.2 to 98 s

		Heat Input, watts								
S. No.	Node	96.2 s	96.4 s	96.6 s	96.8 s	97.0 s	97.2 s	97.4 s	97.6 s	97.8 s	98.0 s
1	630	184.19	149.56	114.39	81.34	52.73	30.47	15.35	6.60	2.38	0.71
2	631	804.70	653.38	499.74	355.36	230.36	133.12	67.06	28.84	10.40	3.10
3	632	184.19	149.56	114.39	81.34	52.73	30.47	15.35	6.60	2.38	0.71
4	643	186.71	222.88	256.88	285.81	305.69	312.82	305.69	285.81	256.88	222.88
5	644	815.72	973.72	1122.24	1248.62	1335.48	1366.64	1335.48	1248.62	1122.24	973.72
6	645	186.71	222.88	256.88	285.81	305.69	312.82	305.69	285.81	256.88	222.88
7	656	0.17	0.71	2.38	6.60	15.35	30.47	52.73	81.34	114.39	149.56
8	657	0.76	3.10	10.40	28.84	67.06	133.12	230.36	355.36	499.74	653.38
9	658	0.17	0.71	2.38	6.60	15.35	30.47	52.73	81.34	114.39	149.56

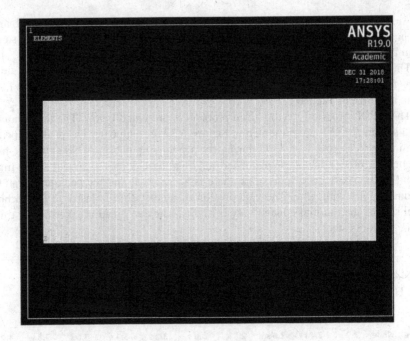

FIGURE 6.37   Discretization of the plate for analysis.

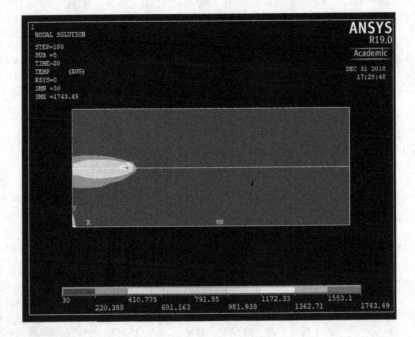

FIGURE 6.38   Temperature distribution in the plate at 20 s.

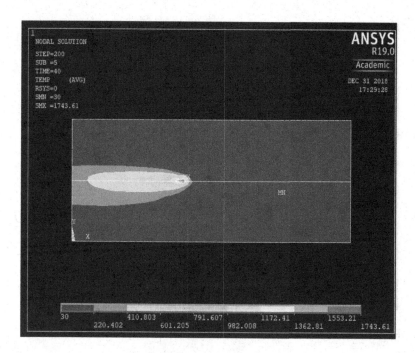

FIGURE 6.39 Temperature distribution in the plate at 40 s.

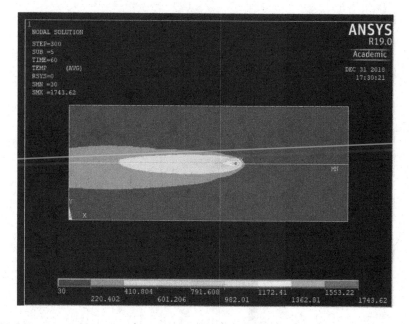

FIGURE 6.40 Temperature distribution in the plate at 60 s.

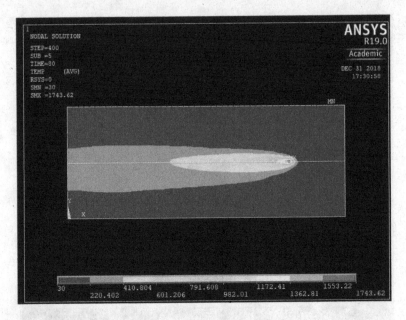

FIGURE 6.41   Temperature distribution in the plate at 80 s.

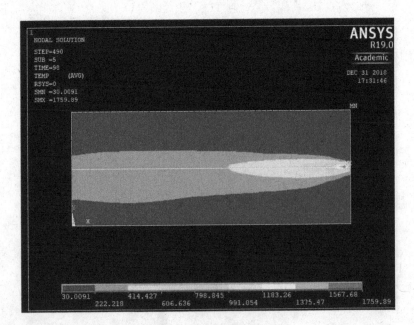

FIGURE 6.42   Temperature distribution in the plate at 98 s.

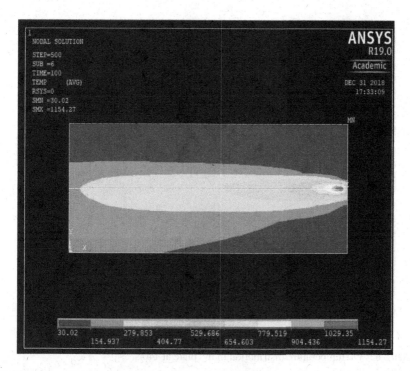

FIGURE 6.43 Temperature distribution in the plate at 100 s.

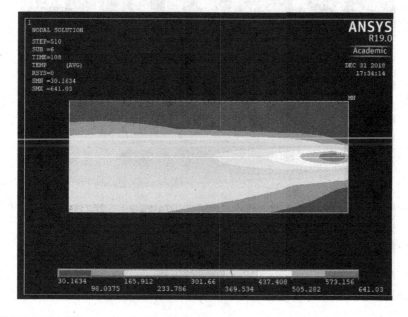

FIGURE 6.44 Temperature distribution in the plate at 108 s.

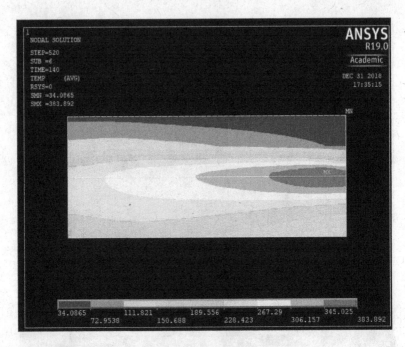

FIGURE 6.45   Temperature distribution in the plate at 140 s.

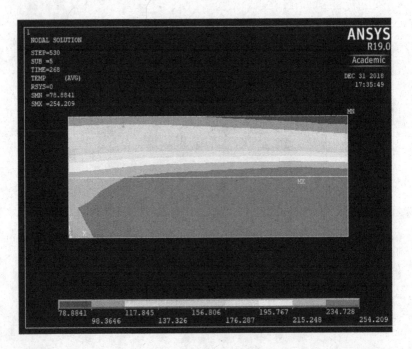

FIGURE 6.46   Temperature distribution in the plate at 268 s.

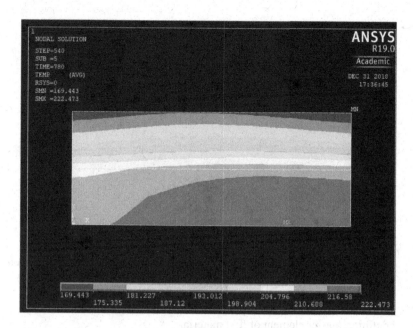

FIGURE 6.47 Temperature distribution in the plate at 780 s.

The APDL program for the case of MIG welded dissimilar weldment

```
/prep7

! defining the element type
et,1,plane55,,,3

! defining the real constant viz the thickness of the plate
r,1,5

! defining the material properties as a function of temperature
! kxx is the thermal conductivity in W/mm K
! c is the specific heat in J / kg K
! dens is the density in kg / mm3
mptemp,1,0,500,1200,1538,1540,3000

! defining for the first material
mpdata,kxx,1,1,0.078,0.058,0.03,0.03,0.06,0.06
mpdata,c,1,1,200,600,600,600,600,600
mpdata,dens,1,1,8e-6,8e-6,8e-6,8e-6,8e-6,8e-6

! defining for the second material
mpdata,kxx,2,1,0.018,0.023,0.03,0.03,0.06,0.06
mpdata,c,2,1,300,600,600,600,600,600
mpdata,dens,2,1,8e-6,8e-6,8e-6,8e-6,8e-6,8e-6
```

```
! defining the master set of nodes
n,1
n,2,,50
n,3,,75
n,4,,85
n,5,,90
n,6,,95
n,7,,100
n,8,,105
n,9,,110
n,10,,115
n,11,,125
n,12,,150
n,13,,200

! generating multiple copies of nodes from the master set using the ngen
command
ngen,51,13,1,13,1,10

! defining master element of first material
mat,1
e,1,14,15,2

! generating multiple copies of elements from the master set using the egen
command
egen,6,1,-1
egen,50,13,-6

! defining the master element of second material
mat,2
e,7,20,21,8

! generating multiple copies of elements from the master set using the egen
command
egen,6,1,-1
egen,50,13,-6
fini
/solu

! defining the analysis type as transient
antype,4
nropt,full
autots,on

! defining uniform temperature and reference temperature
tunif,30
tref,30
```

```
! defining the number of substeps in each time increment and the number
of iterations
nsubst,20
neqit,20
timint,on,ther
tintp,,,,1
eqslv,jcg

! defining convection boundary condition on all nodes
sf,all,conv,2e-6,30
fdele,all,all

! entering in the do loop for first 490 steps
! the arc heat is experienced only during these time steps
*do,i,1,49
*do,j,1,10
n1=(i-1)*13+6
n2=(i-1)*13+7
n3=(i-1)*13+8
n4=n1+13
n5=n2+13
n6=n3+13
n7=n1+26
n8=n2+26
n9=n3+26

! defining the time step
time,(i-1)*2+j*0.2

! defining the nodes which receive the arc heat and the corresponding arc
heat values
*if,j,eq,1,then
f1=184.19
f2=804.70
f3=184.19
f4=186.71
f5=815.72
f6=186.71
f7=0.17
f8=0.76
f9=0.17
*endif
*if,j,eq,2,then
f1=149.56
f2=653.38
f3=149.56
```

f4=222.88
f5=973.72
f6=222.88
f7=0.71
f8=3.10
f9=0.71
*endif
*if,j,eq,3,then
f1=114.39
f2=499.74
f3=114.39
f4=256.88
f5=1122.24
f6=256.88
f7=2.38
f8=10.40
f9=2.38
*endif
*if,j,eq,4,then
f1=81.34
f2=355.36
f3=81.34
f4=285.81
f5=1248.62
f6=285.81
f7=6.60
f8=28.84
f9=6.60
*endif
*if,j,eq,5,then
f1=52.73
f2=230.36
f3=52.73
f4=305.69
f5=1335.48
f6=305.69
f7=15.35
f8=67.06
f9=15.35
*endif
*if,j,eq,6,then
f1=30.47
f2=133.12
f3=30.47

f4=312.82
f5=1366.64
f6=312.82
f7=30.47
f8=133.12
f9=30.47
*endif
*if,j,eq,7,then
f1=15.35
f2=67.06
f3=15.35
f4=305.69
f5=1335.48
f6=305.69
f7=52.73
f8=230.36
f9=52.73
*endif
*if,j,eq,8,then
f1=6.60
f2=28.84
f3=6.60
f4=285.81
f5=1248.62
f6=285.81
f7=81.34
f8=355.36
f9=81.34
*endif
*if,j,eq,9,then
f1=2.38
f2=10.40
f3=2.38
f4=256.88
f5=1122.24
f6=256.88
f7=114.39
f8=499.74
f9=114.39
*endif
*if,j,eq,10,then
f1=0.71
f2=3.10
f3=0.71

```
f4=222.88
f5=973.72
f6=222.88
f7=149.56
f8=653.38
f9=149.56
*endif
fdele,all,all
f,n1,heat,f1
f,n2,heat,f2
f,n3,heat,f3
f,n4,heat,f4
f,n5,heat,f5
f,n6,heat,f6
f,n7,heat,f7
f,n8,heat,f8
f,n9,heat,f9
solve
*enddo
*enddo


! entering in the next set of do loops
! the arc heat is absent during these subsequent time steps
*do,i,1,10
time,98+i*0.2
fdele,all,all
solve
*enddo
*do,i,1,10
time,100+i*0.8
fdele,all,all
solve
*enddo
*do,i,1,10
time,108+i*3.2
fdele,all,all
solve
*enddo
*do,i,1,10
time,140+i*12.8
fdele,all,all
```

```
solve
*enddo
*do,i,1,10
time,268+i*51.2
fdele,all,all
solve
*enddo
*do,i,1,10
time,780+i*204.8
fdele,all,all
solve
*enddo
fini
```

6.4   IN-PLANE ANALYSIS OF A GAS METAL ARC WELDED PLATE

Problem

Perform the in-plane thermal analysis for the following case.

Welding process = Gas metal arc welding
Welding current = 70 A
Welding voltage = 16 V
Process efficiency = 0.786
Welding speed = 4 mm/s
Arc diameter = 8 mm
Plate size = 400 mm long 100 mm wide and 3 mm thick
Element size near the arc zone = 2 × 2 mm
Movement of the arc for each time step = 0.5 mm

Since the plate thickness is only 3 mm, a two-dimensional in-plane type of analysis is suitable for the problem. The symmetry condition has been invoked, hence only a half section is taken for the analysis.

The material properties, thermal conductivity and specific heat are assumed to be temperature dependent as in the earlier case.

The next step is to calculate the heat values which are experienced at the four corner nodes of the element which lies along the weld length. The heat input values have been calculated for the duration of 0 to 2.5 s with a time increment of 0.125 s. The values which are calculated as described in Section 5 (example problem 5.2.4) are given in Table 6.12 and 6.13.

Table 6.12 gives the heat input for the element which has a y coordinate varying from 0 to 2 mm. Table 6.13 gives the heat input values for the element which has a y coordinate from 2 to 4 mm.

In a set of elements which lie along the welding direction as shown in Figure 6.48, the arc is distributed over eight elements at t = 0 s. The arc is about to enter in elements 5 and 10. From the position of the arc, it can be seen that as the arc advances, the heat values from the table are applied to the elements 5, 4, 3, 2, and 1 corresponding to a time step of 0, 0.5, 1.0, 1.5, and 2 s, respectively. After 0.5 s, the arc covers elements 5 and 10 and completely moves out of elements 1 and 6. The arc heat distribution for various positions of the arc is presented in Figure 6.48.

The heat input liberated at various nodes can be added up and the nodal heat input for various time intervals is as shown in Figure 6.49.

TABLE 6.12
Calculated Heat Input for Element 1 with y Ranging from 0 to 2 mm

Time, s	Heat Input, watts			
	Node 1	Node 2	Node 3	Node 4
0	1.01	0.34	0.27	0.80
0.125	2.19	0.81	0.64	1.73
0.25	4.36	1.78	1.40	3.44
0.375	7.92	3.57	2.82	6.25
0.5	13.17	6.61	5.21	10.40
0.625	20.05	11.22	8.85	15.83
0.75	27.95	17.50	13.81	22.06
0.875	35.69	25.06	19.78	28.16
1.0	41.76	32.95	26.00	32.95
1.125	44.79	39.78	31.39	35.35
1.25	44.07	44.07	34.78	34.78
1.375	39.78	44.79	35.35	31.39
1.5	32.95	41.76	32.95	26.00
1.625	25.06	35.69	28.16	19.78
1.75	17.50	27.95	22.06	13.81
1.875	11.22	20.05	15.83	8.85
2.0	6.61	13.17	10.40	5.21
2.125	3.57	7.92	6.25	2.82
2.25	1.78	4.36	3.44	1.40
2.375	0.81	2.19	1.73	0.64
2.5	0.34	1.01	0.80	0.27

TABLE 6.13

Calculated Heat Input for Element 2 with y Ranging from 2 to 4 mm

Time, s	Heat Input, watts			
	Node 1	Node 2	Node 3	Node 4
0	0.32	0.11	0.05	0.16
0.125	0.69	0.26	0.13	0.35
0.25	1.38	0.56	0.28	0.69
0.375	2.50	1.13	0.57	1.25
0.5	4.16	2.08	1.05	2.08
0.625	6.33	3.54	1.78	3.17
0.75	8.82	5.52	2.77	4.42
0.875	11.26	7.91	3.97	5.65
1.0	13.17	10.40	5.21	6.61
1.125	14.13	12.55	6.30	7.09
1.25	13.90	13.90	6.97	6.97
1.375	12.55	14.13	7.09	6.30
1.5	10.40	13.17	6.61	5.21
1.625	7.91	11.26	5.65	3.97
1.75	5.52	8.82	4.42	2.77
1.875	3.54	6.33	3.17	1.78
2.0	2.08	4.16	2.08	1.05
2.125	1.13	2.50	1.25	0.57
2.25	0.56	1.38	0.69	0.28
2.375	0.26	0.69	0.35	0.13
2.5	0.11	0.32	0.16	0.05

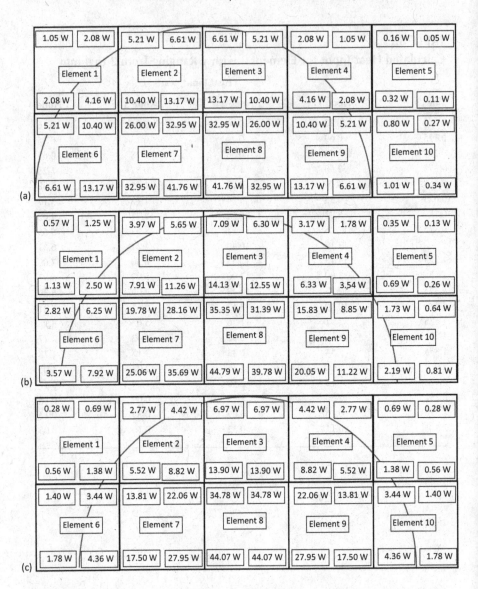

FIGURE 6.48 (a) Heat distribution in various elements corresponding to t = 0 s, (b) heat distribution in various elements corresponding to t = 0.125 s, (c) heat distribution in various elements corresponding to t = 0.25 s. *(Continued)*

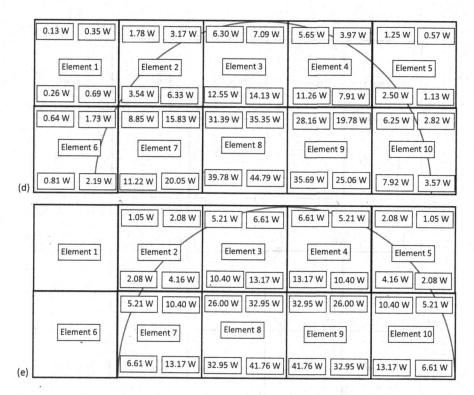

FIGURE 6.48 (Continued) (d) heat distribution in various elements corresponding to t = 0.375 s, and (e) heat distribution in various elements corresponding to t = 0.5 s.

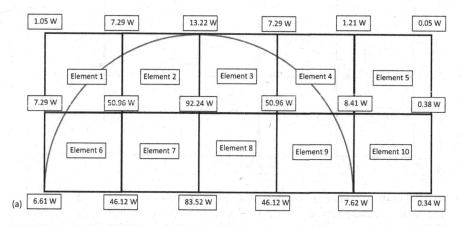

FIGURE 6.49 (a) The nodal heat distribution at time t = 0 s. The total heat input at various nodes adds to 430.68 W.   (*Continued*)

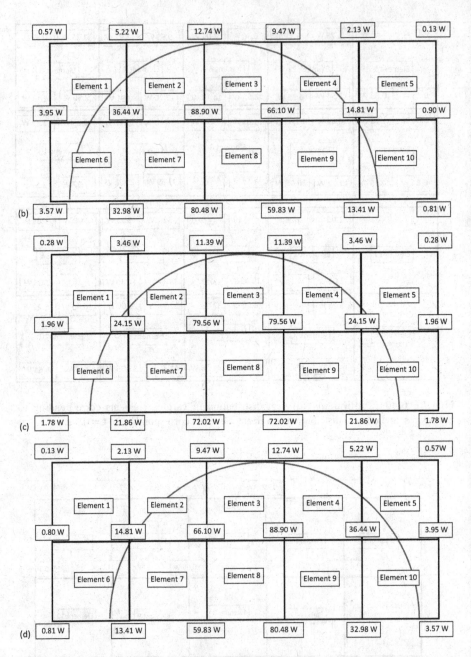

FIGURE 6.49 (Continued) (b) the nodal heat distribution at time t = 0.125 s. The total heat input at various nodes adds to 432.44 W, (c) the nodal heat distribution at time t = 0.25 s. The total heat input at various nodes adds to 432.92 W, and (d) the nodal heat distribution at time t = 0.375 s. The total heat input at various nodes adds to 432.34 W.

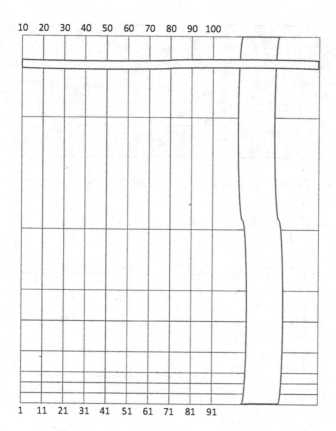

FIGURE 6.50 Discretization of the plate for analysis. Picture is not to scale.

The discretization of the plate into various elements and nodes is performed using direct generation of nodes and elements as shown in Figure 6.50. The length of the plate, which lies along the x direction, is divided into 200 equal segments of 2 mm each. In the y direction, the element size has been selected to be 2 mm near the weld. The spacing is gradually increased, and there are 9 elements in the y direction. Thus, there are a total of $10 \times 201 = 2010$ nodes and 1800 elements in the plate.

Initially, the arc has its centre at a distance of 4 mm from the edge as shown in Figure 6.51. Similarly, during the last step, the arc has its centre at a distance of 4 mm from the other edge. Thus during the arc moving phase, the arc moves at a distance of 392 mm which takes 98 s. After 98 s, there will be no arc heat input in the plate and the cooling phase begins at this time step.

The position of the arc at time t = 0 s is shown in Figure 6.52a. Since the time step in ANSYS cannot begin with a time t = 0, the heat values which have been calculated for a time t = 0 have been assigned for a time step of 0.125 s. The nodal heat values for the time steps from 0 to 0.5 s are shown in Table 6.14.

At 0.5 s, the arc advances to the next set of elements as shown in Figure 6.52b and the heat input values at various nodes are as shown in Table 6.15 for the time steps from 0.625 s to 1.0 s.

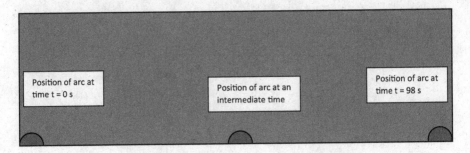

FIGURE 6.51 The starting, middle, and ending position of the arc along the weld line.

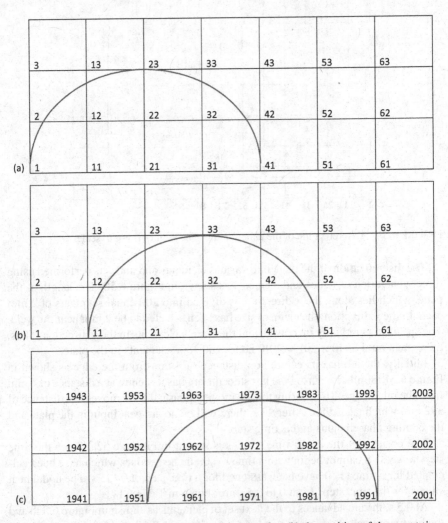

FIGURE 6.52 (a) The position of the arc at time t = 0 s, (b) the position of the arc at time t = 0.5 s, and (c) the position of the arc at time t = 97.5 s.

TABLE 6.14
Nodal Heat Input Corresponding to 0.125 to 0.5 s

S. No.	Node	Heat Input, watts			
		T = 0.125 s	T = 0.25 s	T = 0.375 s	T = 0.5 s
1	1	6.61	3.57	1.78	0.81
2	2	7.29	3.95	1.96	0.8
3	3	1.05	0.57	0.28	0.13
4	11	46.12	32.98	21.86	13.41
5	12	50.96	36.44	24.15	14.81
6	13	7.29	5.22	3.46	2.13
7	21	83.52	80.48	72.02	59.83
8	22	92.24	88.9	79.56	66.1
9	23	13.22	12.74	11.39	9.47
10	31	46.12	59.83	72.02	80.48
11	32	50.96	66.1	79.56	88.9
12	33	7.29	9.47	11.39	12.74
13	41	7.62	13.41	21.86	32.98
14	42	8.41	14.81	24.15	36.44
15	43	1.21	2.13	3.46	5.22
16	51	0.34	0.81	1.78	3.57
17	52	0.38	0.9	1.96	3.95
18	53	0.05	0.13	0.28	0.57

TABLE 6.15
Nodal Heat Input Corresponding to 0.625 to 1.0 s

S. No.	Node	Heat Input, watts			
		T = 0.625 s	T = 0.75 s	T = 0.875 s	T = 1.0 s
1	11	6.61	3.57	1.78	0.81
2	12	7.29	3.95	1.96	0.8
3	13	1.05	0.57	0.28	0.13
4	21	46.12	32.98	21.86	13.41
5	22	50.96	36.44	24.15	14.81
6	23	7.29	5.22	3.46	2.13
7	31	83.52	80.48	72.02	59.83
8	32	92.24	88.9	79.56	66.1
9	33	13.22	12.74	11.39	9.47
10	41	46.12	59.83	72.02	80.48
11	42	50.96	66.1	79.56	88.9
12	43	7.29	9.47	11.39	12.74

(Continued)

TABLE 6.15 (Continued)
Nodal Heat Input Corresponding to 0.625 to 1.0 s

S. No.	Node	Heat Input, watts			
		T = 0.625 s	T = 0.75 s	T = 0.875 s	T = 1.0 s
13	51	7.62	13.41	21.86	32.98
14	52	8.41	14.81	24.15	36.44
15	53	1.21	2.13	3.46	5.22
16	61	0.34	0.81	1.78	3.57
17	62	0.38	0.9	1.96	3.95
18	63	0.05	0.13	0.28	0.57

TABLE 6.16
Nodal Heat Input Corresponding to 97.625 to 98.0 s

S. No.	Node	Heat Input, watts			
		T = 97.625 s	T = 97.75 s	T = 97.875 s	T = 98.0 s
1	1951	6.61	3.57	1.78	0.81
2	1952	7.29	3.95	1.96	0.8
3	1953	1.05	0.57	0.28	0.13
4	1961	46.12	32.98	21.86	13.41
5	1962	50.96	36.44	24.15	14.81
6	1963	7.29	5.22	3.46	2.13
7	1971	83.52	80.48	72.02	59.83
8	1972	92.24	88.9	79.56	66.1
9	1973	13.22	12.74	11.39	9.47
10	1981	46.12	59.83	72.02	80.48
11	1982	50.96	66.1	79.56	88.9
12	1983	7.29	9.47	11.39	12.74
13	1991	7.62	13.41	21.86	32.98
14	1992	8.41	14.81	24.15	36.44
15	1993	1.21	2.13	3.46	5.22
16	2001	0.34	0.81	1.78	3.57
17	2002	0.38	0.9	1.96	3.95
18	2003	0.05	0.13	0.28	0.57

In this way, the heat input values are entered at the nodes. The analysis performed for the last set of elements is as shown in Figure 6.52c. The corresponding heat values are given in Table 6.16.

During the arc movement phase, the time increment is maintained at 0.125 s, and during the cooling phase, the time intervals are progressively increased and the analysis is performed up to a time of 5560 s.

The APDL program for the problem is presented below. The temperature plots corresponding to various time intervals are as shown in Figures 6.53 through 6.61.

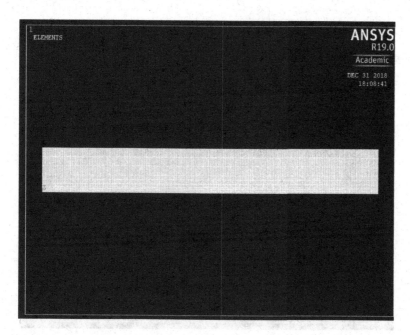

FIGURE 6.53 Discretization of the plate for analysis.

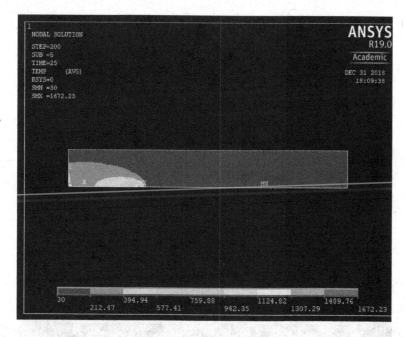

FIGURE 6.54 Temperature distribution in the plate at 25 s.

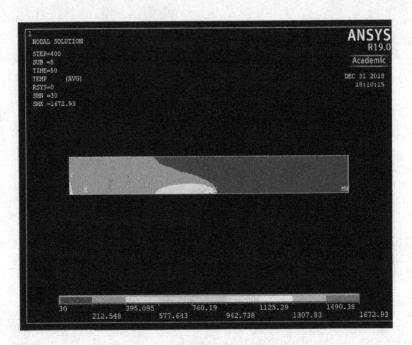

FIGURE 6.55   Temperature distribution in the plate at 50 s.

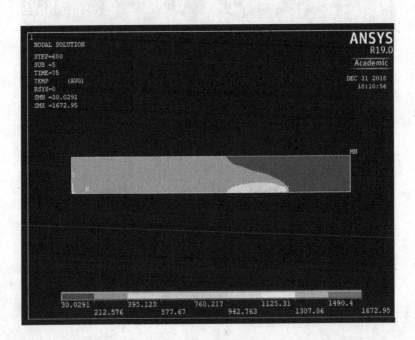

FIGURE 6.56   Temperature distribution in the plate at 75 s.

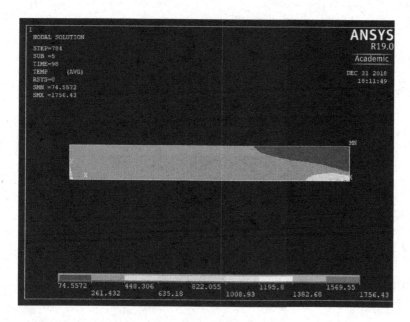

FIGURE 6.57   Temperature distribution in the plate at 98 s.

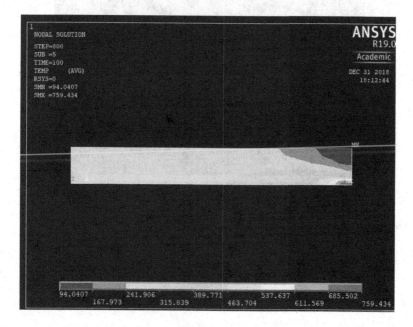

FIGURE 6.58   Temperature distribution in the plate at 100 s.

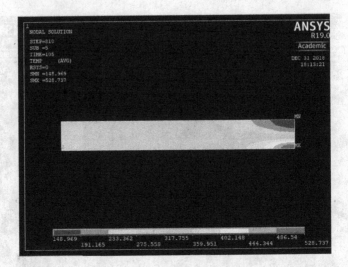

FIGURE 6.59   Temperature distribution in the plate at 105 s.

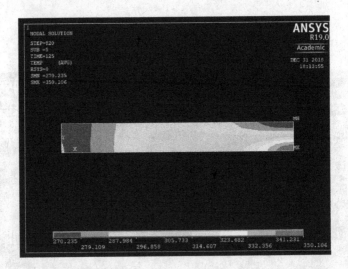

FIGURE 6.60   Temperature distribution in the plate at 125 s.

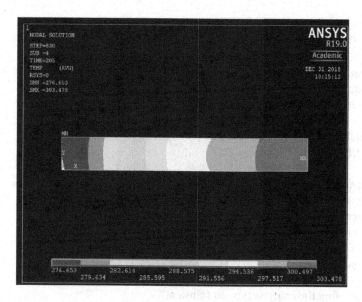

FIGURE 6.61 Temperature distribution in the plate at 205 s.

The APDL code for the case of MIG welded plate

/prep7

! defining the element type
et,1,plane55,,,3

! defining the real constant viz the plate thickness
r,1,3

! defining the material properties as a function of temperature
! kxx is the thermal conductivity in W/mm K
! c is the specific heat in J/kg K
! dens is the density in kg / mm3
mptemp,1,0,500,1200,1538,1540,3000
mpdata,kxx,1,1,0.078,0.058,0.03,0.03,0.06,0.06
mpdata,c,1,1,200,600,600,600,600,600
mpdata,dens,1,1,8e-6,8e-6,8e-6,8e-6,8e-6,8e-6

! defining the master set of nodes
n,1
n,2,,2
n,3,,4
n,4,,6
n,5,,10
n,6,,15
n,7,,20
n,8,,30

```
n,9,,40
n,10,,50
```

```
! generating multiple sets of nodes from the master set using the ngen
command
ngen,201,10,1,10,1,2
```

```
! defining the first element
e,1,11,12,2
```

```
! generating multiple sets of elements from the master set using the egen
command
egen,9,1,-1
egen,200,10,-9
fini
/solu
```

```
! defining the analysis type as transient
antype,4
nropt,full
autots,on
```

```
! defining uniform temperature and reference temperature
tunif,30
tref,30
```

```
! defining the number of substeps in each time increment and the number
of iterations
nsubst,20
neqit,20
timint,on,ther
tintp,,,,1
eqslv,jcg
```

```
! applying convection boundary condition on all nodes
sf,all,conv,5e-6,30
fdele,all,all
```

```
! entering in the first set of do loops
! the arc is moving along the plate during these time steps
*do,i,1,196
```

```
! defining nodes which receive the arc heat
n1=(i-1)*10+1
n2=(i-1)*10+2
n3=(i-1)*10+3
```

```
n4=i*10+1
n5=i*10+2
n6=i*10+3
n7=(i+1)*10+1
n8=(i+1)*10+2
n9=(i+1)*10+3
n10=(i+2)*10+1
n11=(i+2)*10+2
n12=(i+2)*10+3
n13=(i+3)*10+1
n14=(i+3)*10+2
n15=(i+3)*10+3
n16=(i+4)*10+1
n17=(i+4)*10+2
n18=(i+4)*10+3
*do,j,1,4

! defining the time step
time,(i-1)*0.5+j*0.125

! defining the heat input values
*if,j,eq,1,then
f1=6.61
f2=7.29
f3=1.05
f4=46.12
f5=50.96
f6=7.29
f7=83.52
f8=92.24
f9=13.22
f10=46.12
f11=50.96
f12=7.29
f13=7.62
f14=8.41
f15=1.21
f16=0.34
f17=0.38
f18=0.05
*endif
*if,j,eq,2,then
f1=3.57
f2=3.95
f3=0.57
f4=32.98
```

```
f5=36.44
f6=5.22
f7=80.48
f8=88.9
f9=12.74
f10=59.83
f11=66.1
f12=9.47
f13=13.41
f14=14.81
f15=2.13
f16=0.81
f17=0.9
f18=0.13
*endif
*if,j,eq,3,then
f1=1.78
f2=1.96
f3=0.28
f4=21.86
f5=24.15
f6=3.46
f7=72.02
f8=79.56
f9=11.39
f10=72.02
f11=79.56
f12=11.39
f13=21.86
f14=24.15
f15=3.46
f16=1.78
f17=1.96
f18=0.28
*endif
*if,j,eq,4,then
f1=0.81
f2=0.8
f3=0.13
f4=13.41
f5=14.81
f6=2.13
f7=59.83
f8=66.1
f9=9.47
f10=80.48
```

```
f11=88.9
f12=12.74
f13=32.98
f14=36.44
f15=5.22
f16=3.57
f17=3.95
f18=0.57
*endif
fdele,all,all

! applying the heat input values at the appropriate nodes
f,n1,heat,f1
f,n2,heat,f2
f,n3,heat,f3
f,n4,heat,f4
f,n5,heat,f5
f,n6,heat,f6
f,n7,heat,f7
f,n8,heat,f8
f,n9,heat,f9
f,n10,heat,f10
f,n11,heat,f11
f,n12,heat,f12
f,n13,heat,f13
f,n14,heat,f14
f,n15,heat,f15
f,n16,heat,f16
f,n17,heat,f17
f,n18,heat,f18
solve
*enddo
*enddo

! entering in the next set of do loops
! the arc heat is absent during these subsequent time steps
*do,i,1,16
time,98+i*0.125
fdele,all,all
solve
*enddo
*do,i,1,10
time,100+i*0.5
fdele,all,all
solve
*enddo
```

```
*do,i,1,10
time,105+i*2
fdele,all,all
solve
*enddo
*do,i,1,10
time,125+i*8
fdele,all,all
solve
*enddo
*do,i,1,10
time,205+i*32
fdele,all,all
solve
*enddo
*do,i,1,10
time,525+i*128
solve
*enddo
*do,i,1,10
time,1805+i*256
solve
*enddo
fini
```

6.5   THREE-DIMENSIONAL ANALYSIS OF A GAS TUNGSTEN ARC WELDED TUBE

The three-dimensional thermal analysis is similar to the in-plane analysis since the arc movement is over a surface area of the component. Hence, after creating the three-dimensional component, the area over which the arc moves and the corresponding nodes must be identified and the heat input must be introduced at these nodes for various time steps. A sample problem of a tube is given below.

Problem

Perform three-dimensional thermal analysis for the following case.

Welding process = Gas tungsten arc welding
Welding current = 80 A
Welding voltage = 12 V
Process efficiency = 0.6
Welding speed = 1 mm/s
Element size in the arc zone = 5 × 2.5 mm
Arc diameter = 4 mm
Tube = 200 mm long, 76.4 mm OD, and 5 mm wall thickness
Distance moved by the arc during each time step = 0.5 mm

The symmetry condition has been invoked and hence, only a half section is taken for the analysis.

The material properties thermal conductivity and specific heat are assumed to be temperature dependent as in the previous cases.

The next step is to calculate the heat values which are experienced by an element that lies along the weld length. The various heat values have been calculated for the time duration of 0 to 9 s with a time increment of 0.5 s as described in Section 5 (example problem 5.2.3) and are given in Table 6.17.

In a set of elements which lie along the welding direction as shown, the arc is distributed over element 1 and is about to enter in element 2 for a time t = 0 s. From the position of the arc, it can be seen that the corresponding time for element 1 is 5 s. As the arc advances into element 2, the heat values from Table 6.17 are applied to the elements. After 4 s, the arc completely moves into element 2. Figure 6.62 shows the arc heat distribution in the elements for various positions of the arc.

TABLE 6.17
Arc Heat Input Values for Various Time Steps

Time, s	Heat Input, watts			
	Node 1	Node 2	Node 3	Node 4
0	1.48	0.07	0.03	0.52
0.5	6.63	0.44	0.15	2.32
1.0	21.60	1.84	0.64	7.54
1.5	51.66	5.77	2.01	18.04
2.0	92.58	13.92	4.86	32.32
2.5	128.50	27.07	9.45	44.87
3.0	145.11	44.44	15.52	50.67
3.5	141.59	64.33	22.46	49.44
4.0	126.17	85.24	29.76	44.05
4.5	106.25	106.25	37.10	37.10
5.0	85.24	126.17	44.05	29.76
5.5	64.33	141.59	49.44	22.46
6.0	44.44	145.11	50.67	15.52
6.5	27.07	128.50	44.87	9.45
7.0	13.92	92.58	32.32	4.86
7.5	5.77	51.66	18.04	2.01
8.0	1.84	21.60	7.54	0.64
8.5	0.44	6.63	2.32	0.15
9.0	0.07	1.48	0.52	0.03

The heat input liberated at various nodes is added up, and the nodal heat input values for various time intervals are given in Figure 6.63.

The pattern gets repeated after 5 s, and the arc heat values shift to the next set of nodes. Thus, if the node numbers are known, then it is possible to determine the heat input especially when the analysis is carried out with the help of a program.

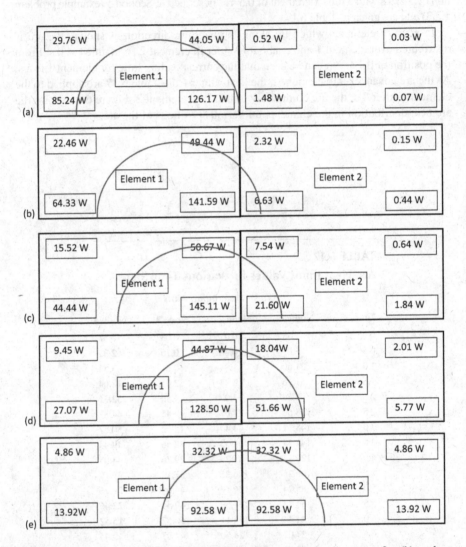

FIGURE 6.62 (a) Arc heat distribution in the elements corresponding to t = 0 s, (b) arc heat distribution in the elements corresponding to t = 0.5 s, (c) arc heat distribution in the elements corresponding to t = 1.0 s, (d) arc heat distribution in the elements corresponding to t = 1.5 s, (e) arc heat distribution in the elements corresponding to t = 2.0 s. *(Continued)*

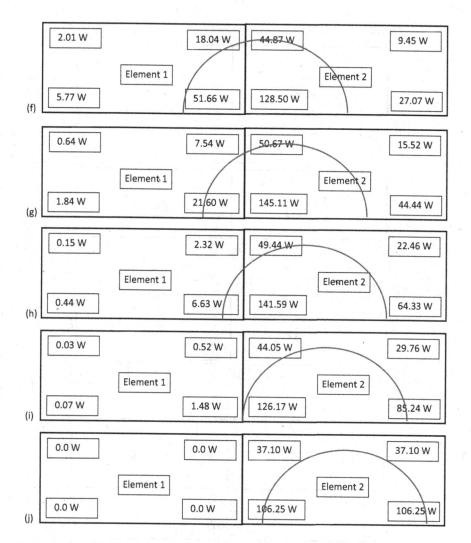

FIGURE 6.62 (Continued) (f) arc heat distribution in the elements corresponding to $t = 2.5$ s, (g) arc heat distribution in the elements corresponding to $t = 3.0$ s, (h) arc heat distribution in the elements corresponding to $t = 3.5$ s, (i) arc heat distribution in the elements corresponding to $t = 4.0$ s, and (j) arc heat distribution in the elements corresponding to $t = 4.5$ s.

The discretization of the tube into various elements and nodes is performed using direct generation of nodes and elements. The outer circumference of the tube over which the arc moves is divided into 48 equal segments of 5 mm each. In the thickness direction, there are four elements and five nodes. The element size near the weld in the axial direction is selected to be 2.5 mm. The spacing is gradually increased,

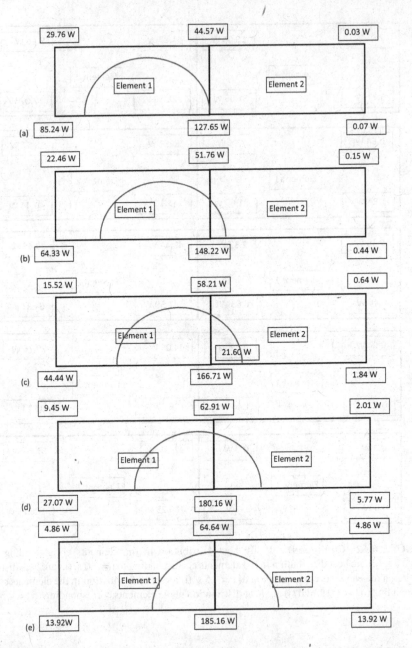

FIGURE 6.63 (a) Arc heat distribution in various nodes corresponding to time t = 0 s. The total heat input at various nodes adds to 287.32 W (b) arc heat distribution in various nodes corresponding to time t = 0.5 s. The total heat input at various nodes adds to 287.36 W (c) arc heat distribution in various nodes corresponding to time t = 1.0 s. The total heat input at various nodes adds to 287.36 W (d) arc heat distribution in various nodes corresponding to time t = 1.5 s. The total heat input at various nodes adds to 287.37 W (e) arc heat distribution in various nodes corresponding to time t = 2.0 s. The total heat input at various nodes adds to 287.36 W.   *(Continued)*

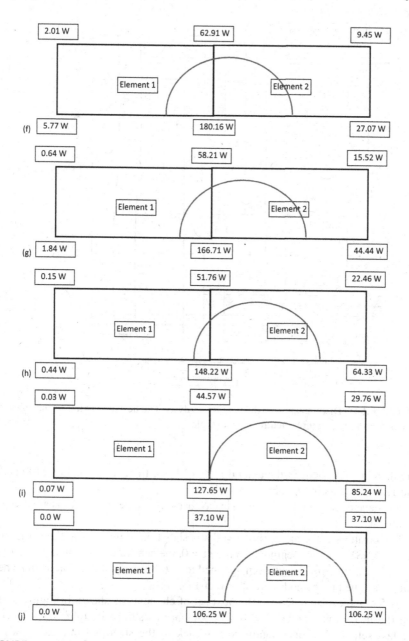

FIGURE 6.63 (Continued)   (f) arc heat distribution in various nodes corresponding to time t = 2.5 s. The total heat input at various nodes adds to 287.37 W (g) arc heat distribution in various nodes corresponding to time t = 3.0 s. The total heat input at various nodes adds to 287.36 W (h) arc heat distribution in various nodes corresponding to time t = 3.5 s. The total heat input at various nodes adds to 287.36 W (i) arc heat distribution in various nodes corresponding to time t = 4.0 ss. The total heat input at various nodes adds to 287.32 W, and (j) arc heat distribution in various nodes corresponding to time t = 4.5 s. The total heat input at various nodes adds to 286.7 W.

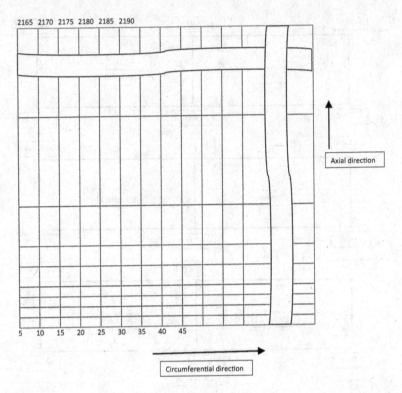

FIGURE 6.64 Discretization of the tube for analysis. The developed view of the outer surface is shown in the figure. Picture is not to scale.

and there are 10 nodes in the axial direction. Thus, there are a total of 2400 nodes and 1728 elements in the tube.

The developed view of the outer surface over which the arc moves is as shown in Figure 6.64.

The position of the arc at time t = 0 s is shown in Figure 6.65a. Since the time step in ANSYS cannot begin with a time t = 0, the heat values which have been calculated for a time t = 0 have been assigned for a time step of 0.5 s. The various heat values assigned to the nodes are shown in Table 6.18.

At 5 s, the arc advances to the next set of elements as shown in Figure 6.65b and the heat input values at various nodes are as shown in Table 6.19. During the last set, the arc once again comes back to the starting point as shown in Figure 6.65c. Table 6.20 gives the heat liberated at various time steps from 235.5 to 240 s.

During arc movement phase, the time increment is maintained at 0.5 s, and during cooling phase, the time intervals are progressively increased and the analysis is performed up to a time of 7065 s.

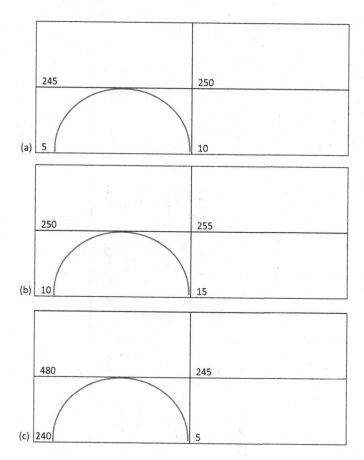

FIGURE 6.65 (a) The position of the arc at time t = 0 s, (b) the position of the arc at time t = 5 s, and (c) the position of the arc at time t = 235 s.

The APDL program for the problem is presented below. The temperature plots corresponding to various time intervals are as shown in Figures 6.66 through 6.71. It is seen that at 240 s, the peak temperature in the weld is 2650°C. But there is a thermal gradient in the thickness direction, and the inside surface is at a much lower temperature. The weld effectively penetrates in the tube to an extent of nearly 1 mm only. The cooling pattern in the tube surface is shown in the other figures.

TABLE 6.18

The Nodal Heat Values for the Time from 0.5 to 5.0 s

S. No.	Node	Heat Input, watts									
		t = 0.5 s	t = 1.0 s	t = 1.5 s	t = 2.0 s	t = 2.5 s	t = 3.0 s	t = 3.5 s	t = 4.0 s	t = 4.5 s	t = 5.0 s
1	5	85.24	64.33	44.44	27.07	13.92	5.77	1.84	0.44	0.07	0.0
2	10	127.65	148.22	166.71	180.16	185.16	180.16	166.71	148.22	127.65	106.25
3	15	0.07	0.44	1.84	5.77	13.92	27.07	44.44	64.33	85.24	106.25
4	245	29.76	22.46	15.52	9.45	4.86	2.01	0.64	0.15	0.03	0
5	250	44.57	51.76	58.21	62.91	64.64	62.91	58.21	51.76	44.57	37.10
6	255	0.03	0.15	0.64	2.01	4.86	9.45	15.52	22.46	29.76	37.10

TABLE 6.19
The Nodal Heat Values for the Time from 5.5 to 10.0 s

S. No.	Node	t = 5.5 s	t = 6.0 s	t = 6.5 s	t = 7.0 s	t = 7.5 s	t = 8.0 s	t = 8.5 s	t = 9.0 s	t = 9.5 s	t = 10.0 s
						Heat Input, watts					
1	10	85.24	64.33	44.44	27.07	13.92	5.77	1.84	0.44	0.07	0.0
2	15	127.65	148.22	166.71	180.16	185.16	180.16	166.71	148.22	127.65	106.25
3	20	0.07	0.44	1.84	5.77	13.92	27.07	44.44	64.33	85.24	106.25
4	250	29.76	22.46	15.52	9.45	4.86	2.01	0.64	0.15	0.03	0
5	255	44.57	51.76	58.21	62.91	64.64	62.91	58.21	51.76	44.57	37.10
6	260	0.03	0.15	0.64	2.01	4.86	9.45	15.52	22.46	29.76	37.10

TABLE 6.20
The Nodal Heat Values for the Time from 235.5 to 240.0 s

S. No.	Node	Heat Input, watts								
		t = 235.5 s	t = 236.0 s	t = 236.5 s	t = 237.0 s	t = 237.5 s	t = 238.0 s	t = 238.5 s	t = 239.0 s	t = 239.5 s	t = 240.0 s
1	240	85.24	64.33	44.44	27.07	13.92	5.77	1.84	0.44	0.07	0.0
2	5	127.65	148.22	166.71	180.16	185.16	180.16	166.71	148.22	127.65	106.25
3	10	0.07	0.44	1.84	5.77	13.92	27.07	44.44	64.33	85.24	106.25
4	480	29.76	22.46	15.52	9.45	4.86	2.01	0.64	0.15	0.03	0
5	245	44.57	51.76	58.21	62.91	64.64	62.91	58.21	51.76	44.57	37.10
6	250	0.03	0.15	0.64	2.01	4.86	9.45	15.52	22.46	29.76	37.10

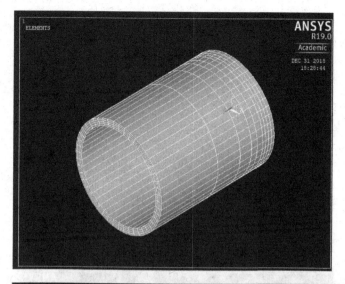

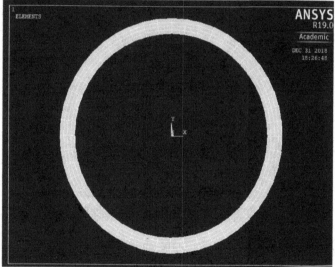

FIGURE 6.66 Discretization of the tube for analysis.

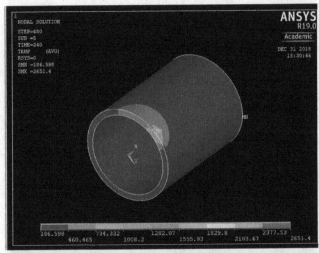

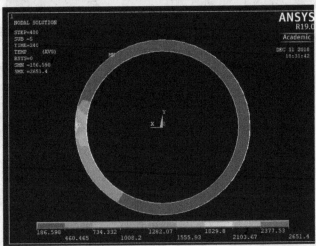

FIGURE 6.67   Temperature distribution in the tube after 240 s.

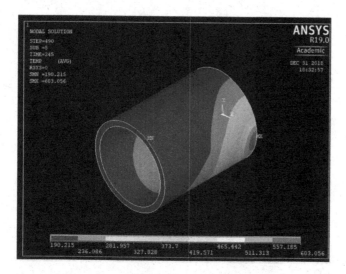

FIGURE 6.68 Temperature distribution in the tube after 245 s.

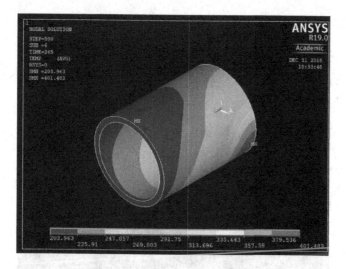

FIGURE 6.69 Temperature distribution in the tube after 265 s.

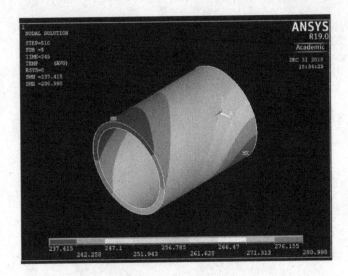

FIGURE 6.70   Temperature distribution in the tube after 345 s.

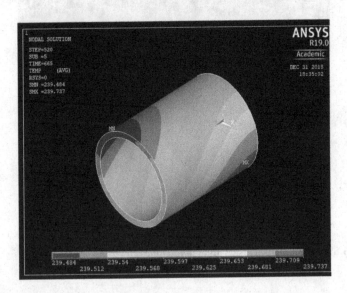

FIGURE 6.71   Temperature distribution in the tube after 665 s.

The APDL program for the three-dimensional analysis of a TIG welded tube

```
/prep7

! defining the element type
et,1,solid278

! defining the material properties as a function of temperature
! kxx is the thermal conductivity in W/mm K
! c is the specific heat in J / kg K
! dens is the density in kg / mm³
mptemp,1,0,500,1200,1538,1540,3000
mpdata,kxx,1,1,0.078,0.058,0.03,0.03,0.06,0.06
mpdata,c,1,1,200,600,600,600,600,600
mpdata,dens,1,1,8e-6,8e-6,8e-6,8e-6,8e-6,8e-6

! changing into cylindrical coordinate system
csys,1

! defining the first set of nodes in radial direction
n,1,33.2
n,5,38.2
fill

! generating multiple copies of nodes from the master set using the ngen command
! the nodes are generated in the circumferential direction
ngen,48,5,1,5,1,,7.5

! generating nodes in axial direction
ngen,5,240,1,240,1,,,2.5
ngen,3,240,961,1200,1,,,5
ngen,2,240,1441,1680,1,,,10
ngen,2,240,1681,1920,1,,,20
ngen,2,240,1921,2160,1,,,50

! defining the first element
e,1,2,7,6,241,242,247,246

! generating multiple copies of elements from the master set using the egen
command
egen,4,1,-1
egen,47,5,-4
e,236,237,2,1,476,477,242,241
egen,4,1,-1
egen,9,240,-192
fini
/solu
```

```
! defining the analysis type as transient
antype,4
nropt,full
autots,on

! defining the uniform temperature and the reference temperature
tunif,30
tref,30

! defining the number of substeps in each time step and number of iterations
nsubst,20
neqit,20
timint,on,ther
tintp,,,,1
eqslv,jcg

! defining the convection boundary condition on all surface nodes
csys,1
nsel,x,38.2
sf,all,conv,2e-6,30
nall
nsel,x,33.2
sf,all,conv,2e-6,30
nall
fdele,all,all

! entering the do loop
! the arc is moving along the tube during these time steps
*do,i,1,48

! defining the nodes which receive arc heat
n1=5*i
n2=n1+240
n3=5*(i+1)
n4=n3+240
n5=5*(i+2)
n6=n5+240
*if,i,eq,47,then
n1=235
n2=475
n3=240
n4=480
n5=5
n6=245
*endif
*if,i,eq,48,then
```

```
n1=240
n2=480
n3=5
n4=245
n5=10
n6=250
*endif
*do,j,1,10

! defining the time step
time,(i-1)*5+j*0.5

! defining the arc heat input
*if,j,eq,1,then
f1=85.24
f2=29.76
f3=127.65
f4=44.57
f5=0.07
f6=0.03
*endif
*if,j,eq,2,then
f1=64.33
f2=22.46
f3=148.22
f4=51.76
f5=0.44
f6=0.15
*endif
*if,j,eq,3,then
f1=44.44
f2=15.52
f3=166.71
f4=58.21
f5=1.84
f6=0.64
*endif
*if,j,eq,4,then
f1=27.07
f2=9.45
f3=180.16
f4=62.91
f5=5.77
f6=2.01
*endif
*if,j,eq,5,then
```

```
f1=13.92
f2=4.86
f3=185.16
f4=64.64
f5=13.92
f6=4.86
*endif
*if,j,eq,6,then
f1=5.77
f2=2.01
f3=180.16
f4=62.91
f5=27.07
f6=9.45
*endif
*if,j,eq,7,then
f1=1.84
f2=0.64
f3=166.71
f4=58.21
f5=44.44
f6=15.52
*endif
*if,j,eq,8,then
f1=0.44
f2=0.15
f3=148.22
f4=51.76
f5=64.33
f6=22.46
*endif
*if,j,eq,9,then
f1=0.07
f2=0.03
f3=127.65
f4=44.57
f5=85.24
f6=29.76
*endif
*if,j,eq,10,then
f1=0
f2=0
f3=106.25
f4=37.10
f5=106.25
f6=37.10
```

```
*endif
fdele,all,all

! applying the arc heat input at the appropriate nodes
f,n1,heat,f1
f,n2,heat,f2
f,n3,heat,f3
f,n4,heat,f4
f,n5,heat,f5
f,n6,heat,f6
solve
*enddo
*enddo

! entering in the next set of do loops
! the arc heat is absent during these time steps
*do,i,1,10
time,240+i*0.5
fdele,all,all
solve
*enddo
*do,i,1,10
time,245+i*2
fdele,all,all
solve
*enddo
*do,i,1,10
time,265+i*8
fdele,all,all
solve
*enddo
*do,i,1,10
time,345+i*32
fdele,all,all
solve
*enddo
*do,i,1,10
time,665+i*128
fdele,all,all
solve
*enddo
*do,i,1,10
time,1945+i*512
fdele,all,all
solve
*enddo
fini
```

7 Conclusion

The finite element method (FEM) is a powerful tool for the analysis of engineering problems, but just like any other tool, the effectiveness of FEM lies with the user. In the hand of an expert, FEM can throw light on many intricate issues, but if the user lacks good understanding of the basics, then the analysis may be totally unreliable. Quite often it is seen that researchers tend to blame FEM for not providing a ready-made solution to their problems without realizing that correct data input is essential for getting meaningful results. Generating a data input for the program may require some effort on the part of the researcher, and if the researcher is reluctant to put in the required effort, then the whole exercise may turn out to be a waste of time. Blaming FEM for not providing good results is like blaming the pen for one's spelling mistakes as FEM is only a tool and cannot generate knowledge on its own.

The key to getting reliable results in welding lies with the proper generation of input data for the FEM program which represents the various physical phenomena accurately. This requires a good understanding of the phenomena as well as the mathematical techniques to express the phenomena as input data. There are several ways in which mistakes can creep into the analysis, and the correctness of the data has to be ascertained at various stages for getting satisfactory results.

This book attempts to give young researchers a step by step procedure for the conduct of transient thermal analysis of the arc welding process. It is hoped that the inputs given in the book are useful for developing the "know how" and "know why" of the analysis.

Appendix
Exercise Problems

1. Calculate the nodal arc heat values for various time intervals for the following case.

Welding process = Gas metal arc welding
Current = 90 A
Voltage = 20 V
Process efficiency = 0.8
Arc diameter = 8 mm
Welding speed = 3 mm/s
Element size = 1 mm
Distance moved by the arc for each time step = 0.75 mm

SOLUTION

Since the element size is 1 mm, there will be four elements in the line of arc movement as shown in Figure A.1.

The calculated nodal arc heat values for the four elements are given in Tables A.1 through A.4.

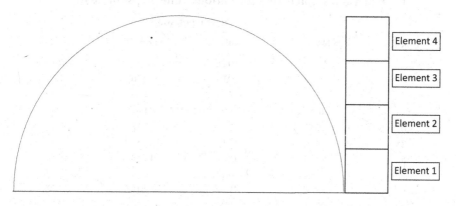

FIGURE A.1   The position of the arc when it is about to enter in the set of elements (t = 0).

TABLE A.1
Arc Heat Values Calculated for Various Time Steps for Element 1

Time, s	Heat Input, watts			
	Node 1	Node 2	Node 3	Node 4
0	0.65	0.38	0.36	0.61
0.25	1.95	1.23	1.16	1.83
0.5	4.72	3.27	3.08	4.44
0.75	9.30	7.05	6.63	8.74
1.0	14.89	12.38	11.64	14.00
1.25	19.38	17.67	16.61	18.22
1.5	20.52	20.52	19.29	19.29
1.75	17.67	19.38	18.22	16.61
2.0	12.38	14.89	14.00	11.64
2.25	7.05	9.30	8.74	6.63
2.5	3.27	4.72	4.44	3.08
2.75	1.23	1.95	1.83	1.16
3.0	0.38	0.65	0.61	0.36

TABLE A.2
Arc Heat Values Calculated for Various Time Steps for Element 2

Time, s	Heat Input, watts			
	Node 1	Node 2	Node 3	Node 4
0	0.48	0.28	0.23	0.40
0.25	1.44	0.91	0.76	1.19
0.5	3.48	2.41	2.00	2.89
0.75	6.85	5.20	4.32	5.69
1.0	10.97	9.12	7.58	9.12
1.25	14.28	13.01	10.82	11.87
1.5	15.11	15.11	12.56	12.56
1.75	13.01	14.28	11.87	10.82
2.0	9.12	10.97	9.12	7.58
2.25	5.20	6.85	5.69	4.32
2.5	2.41	3.48	2.89	2.00
2.75	0.91	1.44	1.19	0.76
3.0	0.28	0.48	0.40	0.23

TABLE A.3

Arc Heat Values Calculated for Various Time Steps for Element 3

Time, s	Heat Input, watts			
	Node 1	Node 2	Node 3	Node 4
0	0.25	0.14	0.10	0.18
0.25	0.73	0.46	0.34	0.54
0.5	1.77	1.23	0.90	1.30
0.75	3.49	2.65	1.95	2.57
1.0	5.59	4.65	3.42	4.11
1.25	7.28	6.64	4.88	5.36
1.5	7.71	7.71	5.67	5.67
1.75	6.64	7.28	5.36	4.88
2.0	4.65	5.59	4.11	3.42
2.25	2.65	3.49	2.57	1.95
2.5	1.23	1.77	1.30	0.90
2.75	0.46	0.73	0.54	0.34
3.0	0.14	0.25	0.18	0.10

TABLE A.4

Arc Heat Values Calculated for Various Time Steps for Element 4

Time, s	Heat Input, watts			
	Node 1	Node 2	Node 3	Node 4
0	0.09	0.05	0.03	0.06
0.25	0.26	0.16	0.11	0.17
0.5	0.63	0.43	0.28	0.41
0.75	1.23	0.93	0.61	0.80
1.0	1.97	1.64	1.07	1.29
1.25	2.57	2.34	1.53	1.68
1.5	2.72	2.72	1.77	1.77
1.75	2.34	2.57	1.68	1.53
2.0	1.64	1.97	1.29	1.07
2.25	0.93	1.23	0.80	0.61
2.5	0.43	0.63	0.41	0.28
2.75	0.16	0.26	0.17	0.11
3.0	0.05	0.09	0.06	0.03

2. Calculate the nodal arc heat values for the following case.

Welding process = Gas tungsten arc welding
Current = 70 A
Voltage = 12 V
Welding speed = 0.8 mm/s
Process efficiency = 0.6
Arc diameter = 5 mm
Element size = 2 mm
Movement of the arc for each time step = 1 mm

SOLUTION

In this case, the arc will be covering two elements along its path as shown in Figure A.2.

The calculated nodal arc heat values for the two elements are given in Tables A.5 and A.6.

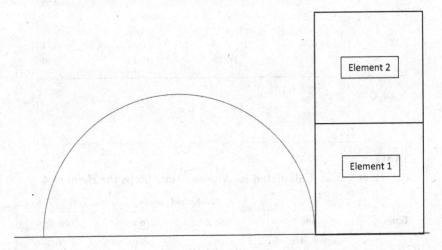

FIGURE A.2   The position of the arc when it is about to enter in the set of elements (t = 0).

TABLE A.5

Arc Heat Values Calculated for Various Time Steps for Element 1

Time, s	Node 1	Node 2	Node 3	Node 4
		Heat Input, watts		
0	0.91	0.18	0.10	0.52
1.25	8.33	2.38	1.37	4.80
2.50	32.00	14.30	8.25	18.45
3.75	53.35	40.34	23.27	30.77
5.00	40.34	53.35	30.77	23.27
6.25	14.30	32.00	18.45	8.25
7.50	2.38	8.33	4.80	1.37
8.75	0.18	0.91	0.52	0.10

TABLE A.6

Arc Heat Values Calculated for Various Time Steps for Element 2

Time, s	Heat Input, watts			
	Node 1	Node 2	Node 3	Node 4
0	0.06	0.01	0.00	0.01
1.25	0.56	0.16	0.04	0.13
2.50	2.15	0.96	0.23	0.51
3.75	3.58	2.71	0.64	0.85
5.00	2.71	3.58	0.85	0.64
6.25	0.96	2.15	0.51	0.23
7.50	0.16	0.56	0.13	0.04
8.75	0.01	0.06	0.01	0.00

3. Calculate the nodal arc heat values for the following case.

Welding process = Submerged arc welding
Current = 300 A
Voltage = 32 V
Process efficiency = 0.9
Welding speed = 10 mm/s
Arc diameter = 8 mm
Element size = 4 × 3 mm
Distance moved by the arc for each time step = 1 mm

SOLUTION

In this case, the arc will be covering two elements along its path as shown in Figure A.3.

The calculated nodal arc heat values for the two elements are given in Tables A.7 and A.8, respectively.

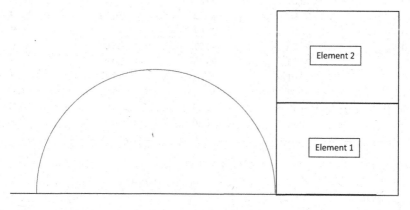

FIGURE A.3 The position of the arc when it is about to enter in the set of elements (t = 0).

TABLE A.7
Arc Heat Values Calculated for Various Time Steps for Element 1

Time, s	Heat Input, watts			
	Node 1	Node 2	Node 3	Node 4
0	15.54	2.39	1.46	9.49
0.1	69.63	13.25	8.09	42.51
0.2	221.87	54.18	33.08	135.45
0.3	507.90	165.98	101.33	310.08
0.4	846.82	387.67	236.67	516.99
0.5	1046.21	699.08	426.79	638.72
0.6	976.06	976.06	595.89	595.89
0.7	699.08	1046.21	638.72	426.79
0.8	387.67	846.82	516.99	236.67
0.9	165.98	507.90	310.08	101.33
1.0	54.18	221.87	135.45	33.08
1.1	13.25	69.63	42.51	8.09
1.2	2.39	15.54	9.49	1.46

TABLE A.8
Arc Heat Values Calculated for Various Time Steps for Element 2

Time, s	Heat Input, watts			
	Node 1	Node 2	Node 3	Node 4
0	1.40	0.21	0.06	0.37
0.1	6.25	1.19	0.32	1.67
0.2	19.93	4.87	1.30	5.31
0.3	45.62	14.91	3.97	12.15
0.4	76.06	34.82	9.28	20.26
0.5	93.97	62.79	16.73	25.03
0.6	87.67	87.67	23.35	23.35
0.7	62.79	93.97	25.03	16.73
0.8	34.82	76.06	20.26	9.28
0.9	14.91	45.62	12.15	3.97
1.0	4.87	19.93	5.31	1.30
1.1	1.19	6.25	1.67	0.32
1.2	0.21	1.40	0.37	0.06

4. Calculate the nodal arc heat values for various time intervals for the following case.

Welding process = Gas metal arc welding
Current = 120 A
Voltage = 22 V
Process efficiency = 0.8
Arc diameter = 8 mm
Welding speed = 5 mm/s
Element size = 1 mm and 2 mm
Distance moved by the arc for each time step = 1 mm

SOLUTION

In this case, there will be three elements of unequal sizes in the path of the arc as shown in Figure A.4.

Tables A.9 through A.11 give the calculated nodal arc heat values for the three elements as given below.

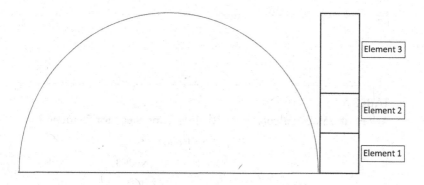

FIGURE A.4   The position of the arc when it is about to enter in the set of elements (t = 0).

TABLE A.9
Arc Heat Values Calculated for Various Time Steps for Element 1

	Heat Input, watts			
Time, s	Node 1	Node 2	Node 3	Node 4
0	0.96	0.56	0.52	0.90
0.2	3.93	2.56	2.41	3.69
0.4	11.13	8.19	7.70	10.47
0.6	21.83	18.15	17.07	20.53
0.8	29.64	27.87	26.20	27.87
1.0	27.87	29.64	27.87	26.20
1.2	18.15	21.83	20.53	17.07
1.4	8.19	11.13	10.47	7.70
1.6	2.56	3.93	3.69	2.41
1.8	0.56	0.96	0.90	0.52

TABLE A.10
Arc Heat Values Calculated for Various Time Steps for Element 2

Time, s	Heat Input, watts			
	Node 1	Node 2	Node 3	Node 4
0	0.71	0.41	0.34	0.59
0.2	2.89	1.89	1.57	2.41
0.4	8.20	6.03	5.02	6.82
0.6	16.08	13.37	11.12	13.37
0.8	21.83	20.53	17.07	18.15
1.0	20.53	21.83	18.15	17.07
1.2	13.37	16.08	13.37	11.12
1.4	6.03	8.20	6.82	5.02
1.6	1.89	2.89	2.41	1.57
1.8	0.41	0.71	0.59	0.34

TABLE A.11
Arc Heat Values Calculated for Various Time Steps for Element 3

Time, s	Heat Input, watts			
	Node 1	Node 2	Node 3	Node 4
0	0.56	0.32	0.16	0.28
0.2	2.28	1.49	0.75	1.14
0.4	6.46	4.75	2.38	3.24
0.6	12.66	10.53	5.28	6.35
0.8	17.19	16.16	8.11	8.62
1.0	16.16	17.19	8.62	8.11
1.2	10.53	12.66	6.35	5.28
1.4	4.75	6.46	3.24	2.38
1.6	1.49	2.28	1.14	0.75
1.8	0.32	0.56	0.28	0.16

5. Calculate the nodal arc heat values for the following case.

Welding process = Gas tungsten arc welding
Current = 40 A
Voltage = 10 V
Process efficiency = 50%
Arc diameter = 4 mm
Welding speed = 1 mm/s
Element size = 10 mm
Distance moved by the arc for each time step = 0.5 mm

SOLUTION

The position of the arc is shown in Figure A.5.
The calculated nodal arc heat values are given in Table A.12.

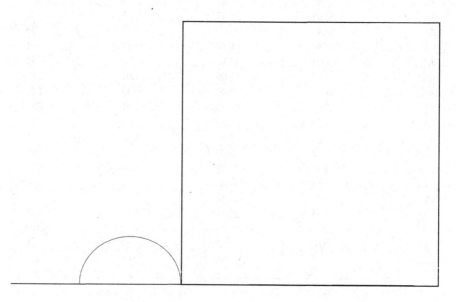

FIGURE A.5   The position of the arc when it is about to enter in the element (t = 0).

TABLE A.12
Arc Heat Values Calculated for Various Time Steps

Time, s	Heat Input, watts			
	Node 1	Node 2	Node 3	Node 4
0	0.80	0.01	0.0	0.06
0.5	3.22	0.06	0.00	0.24
1.0	9.65	0.30	0.02	0.73
1.5	22.30	1.15	0.09	1.68
2.0	41.10	3.11	0.23	3.11
2.5	61.07	6.15	0.46	4.61
3.0	73.38	9.59	0.72	5.54
3.5	74.20	13.18	1.00	5.61
4.0	69.92	17.69	1.34	5.28
4.5	67.39	23.14	1.75	5.09
5.0	65.03	27.88	2.11	4.91
5.5	59.40	31.26	2.36	4.49
6.0	53.05	35.37	2.67	4.01
6.5	49.40	41.27	3.12	3.73
7.0	46.46	46.46	3.51	3.51
7.5	41.27	49.40	3.73	3.12
8.0	35.37	53.05	4.01	2.67
8.5	31.26	59.40	4.49	2.36
9.0	27.88	65.03	4.91	2.11
9.5	23.14	67.39	5.09	1.75
10.0	17.69	69.92	5.28	1.34
10.5	13.18	74.20	5.61	1.00
11.0	9.59	73.38	5.54	0.72
11.5	6.15	61.07	4.61	0.46
12.0	3.11	41.10	3.11	0.23
12.5	1.15	22.30	1.68	0.09
13.0	0.30	9.65	0.73	0.02
13.5	0.06	3.22	0.24	0.00
14.0	0.01	0.80	0.06	0.00

6. **Calculate the nodal arc heat values for the following case.**
   Welding process = Gas tungsten arc welding
   Current = 50 A
   Voltage = 12 V
   Process efficiency = 65%
   Arc diameter = 4 mm
   Welding speed = 2 mm/s
   Element size = 2 mm
   Distance moved by the arc for each time step = 0.1 mm

SOLUTION

The position of the arc is shown in Figure A.6.
The calculated nodal arc heat values are given in Table A.13.

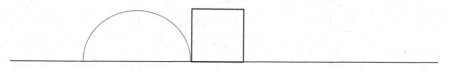

FIGURE A.6   The position of the arc when it is about to enter in the element (t = 0).

TABLE A.13
Arc Heat Values Calculated for Various Time Steps

Time, s	Heat Input, watts			
	Node 1	Node 2	Node 3	Node 4
0	0.82	0.13	0.06	0.37
0.05	1.13	0.18	0.08	0.52
0.1	1.55	0.26	0.12	0.71
0.15	2.10	0.36	0.17	0.96
0.2	2.79	0.51	0.23	1.28
0.25	3.67	0.70	0.32	1.68
0.3	4.75	0.95	0.43	2.17
0.35	6.07	1.27	0.58	2.78
0.4	7.65	1.68	0.77	3.50
0.45	9.51	2.20	1.01	4.36
0.5	11.68	2.85	1.31	5.35
0.55	14.15	3.65	1.67	6.48
0.6	16.91	4.62	2.11	7.74
0.65	19.96	5.77	2.64	9.14
0.7	23.25	7.14	3.27	10.64
0.75	26.73	8.74	4.00	12.24
0.8	30.36	10.57	4.84	13.90
0.85	34.04	12.66	5.80	15.58
0.9	37.70	15.00	6.87	17.26
0.95	41.24	17.59	8.05	18.88
1.00	44.58	20.41	9.34	20.41
1.05	47.60	23.43	10.73	21.79
1.1	50.23	26.64	12.19	23.00
1.15	52.39	29.97	13.72	23.99
1.2	54.02	33.38	15.28	24.73
1.25	55.07	36.80	16.85	25.21
1.3	55.51	40.16	18.38	25.41

(Continued)

TABLE A.13 (*Continued*)

Arc Heat Values Calculated for Various Time Steps

Time, s	Heat Input, watts			
	Node 1	Node 2	Node 3	Node 4
1.35	55.34	43.38	19.86	25.33
1.4	54.57	46.37	21.23	24.98
1.45	53.23	49.07	22.46	24.37
1.5	51.38	51.38	23.52	23.52
1.55	49.07	53.23	24.37	22.46
1.6	46.37	54.57	24.98	21.23
1.65	43.38	55.34	25.33	19.86
1.7	40.16	55.51	25.41	18.38
1.75	36.80	55.07	25.21	16.85
1.8	33.38	54.02	24.73	15.28
1.85	29.97	52.39	23.99	13.72
1.9	26.64	50.23	23.00	12.19
1.95	23.43	47.60	21.79	10.73
2.00	20.41	44.58	20.41	9.34
2.05	17.59	41.24	18.88	8.05
2.1	15.00	37.70	17.26	6.87
2.15	12.66	34.04	15.58	5.80
2.2	10.57	30.36	13.90	4.84
2.25	8.74	26.73	12.24	4.00
2.3	7.14	23.25	10.64	3.27
2.35	5.77	19.96	9.14	2.64
2.4	4.62	16.91	7.74	2.11
2.45	3.65	14.15	6.48	1.67
2.5	2.85	11.68	5.35	1.31
2.55	2.20	9.51	4.36	1.01
2.6	1.68	7.65	3.50	0.77
2.65	1.27	6.07	2.78	0.58
2.7	0.95	4.75	2.17	0.43
2.75	0.70	3.67	1.68	0.32
2.8	0.51	2.79	1.28	0.23
2.85	0.36	2.10	0.96	0.17
2.9	0.26	1.55	0.71	0.12
2.95	0.18	1.13	0.52	0.08
3.00	0.13	0.82	0.37	0.06

Bibliography

1. American Welding Society. 1991. *Welding Handbook*. 8th edition. American Welding Society Publication, Miami, FL.
2. Masubuchi K. 1980. *Analysis of Welded Structures*. 1st edition. Pergamon Press, Oxford.
3. Radaj D. 1988. *Heat Effects of Welding*. Springer Verlag, Berlin, Germany.
4. Segerlind L J. 1976. *Applied Finite Element Analysis*. John Wiley & Sons, New York.
5. Goldak J, Bibby M, Moore J, House R and Patel B. 1986. Computer modeling of heat flow in welds. *Metallurgical Transactions B*, Vol 17B, pp. 587–600.
6. Goldak J, McDill M, Oddy A, House R, Chi X and Bibby M. 1986. Computational heat transfer for weld mechanics. *Proceedings of International Conference on Trends in Welding Research*, USA. Ed David S A. ASM International, pp. 15–20.
7. Friedman E. 1975. Thermomechanical analysis of welding process using finite element method. *Transactions of ASME Journal of Pressure Vessel Technology*, Vol 97, pp. 206–213.
8. Krutz G W and Segerlind L J. 1978. Finite element analysis of welded structures. *Welding Journal*, Vol 57 (7), pp. 211s–216s.
9. ANSYS – Modeling and meshing guide. 2018. ANSYS release 9.0, 2004. ANSYS, Inc. PA.
10. ANSYS – Thermal analysis guide. 2018. ANSYS release 9.0, 2004. ANSYS, Inc. PA.
11. ANSYS – Command reference. 2018.
12. ANSYS – ANSYS Parametric Design Language guide. 2018.

Index

Note: Page numbers in italic and bold refer to figures and tables, respectively.

A

abstraction of heat, 18
aftereffects, 2–3, 17
aluminium alloy, 2
animation, 32
ANSYS, 3, 32–33, 109–110, *111–118*, 132, *134–139*, 152–153, *156–161*, 173, *177–181*, 192, *197–200*
APDL, 32–33, 109, 111, 132, 155, 176, 193
arc, 2, 5–8, 10–15, 17, 19–23, 35–41, *42–43*, 45, 48–55, 69–71, 86, 90, *93*, 109–111, 126–130, 132, *133*, 144–145, *146*, 147, 152–153, *154*, 155, 167–168, 173, *174*, 176, 186–187, 189, 192, *193*, 209, *210*, *212*, 213, *214*, 215, *217*, 218, *219*
   diameter, 10–11, 35, 45, 50–51, 54–55, 85, 90, 126, 129, 144, 152, 167, 186, 209, 213, 215, 217–218
   heat, 6–8, 11, 17, 22, 32–33, 35–42, *43*, 45, **49**, 50–52, 55, 69–71, **84**, *85–86*, **90**, *91–92*, **93–101**, *102–107*, 109–110, 127, *128*, 129–130, 132, 145, 152, **154**, 155, 168, 173, 187, *188–191*, 209, **210–212**, 213, **214**, 215, **216**, 217–218, **219–220**
   length, 6–9, 11, 15, 35, *36*
   radius, 39, 42, 45
   welding, 1–2, 5–7, 11–17, 45, 55, 71, 85, 90, 126, 144, 167, 186, 207, 209, 215, 217
argon, 13
automatic mesh generation, 32
automatic voltage correction, 8, 15
automatic welding, 7–9, 12, 15, 35
automation, 1

B

backhand technique, 10, 35, *36*
base material, 1
base metal, 2, 5–8, 10–15, 17–19, 21, 35–36
basic coated electrode, 12
beam energy, 5
beam welding process, 35
blanketing action, 13
boundary condition, 25, 30, 110

C

carbon steel, 2, 144, 155
cellulose coated electrode, 12
chemical energy, 5
chemical reaction, 11
circular, 35–37, 69
coalescence, 1, 17
computation, 3, 45, 69
conduction, 17–19
constant current characteristics, *9*
constant voltage characteristics, 8, *9*
consumable electrode, 6–8
convection, 6, 12–13, 17–18, 22, 30, 32, 110
   coefficient, 18, 30, 110
convergence, 31, 44, 54
cooling phase, 2, 12, 50, 132, 152, 173, 176, 192
coordinate, 25–28, 32, *37*, 38–39, *93*, 168
cross section, 19, 23, 40–41, 109, **110**
cross sectional analysis, 39, 41, 109, *111*, *115*
cross sectional model, 19, 23
current, 5–9, 11–17, 22, 31, 37, 45, 51, 55, 69, 71, 85, 90, 109, 126, 144, 152, 167, 186, 209, 215, 217
cushioning, 10

D

degradation, 3
density, 6, 11, 22, 29, 32, 35, 109, **127**, 145
differential thermal expansion, 2
digging, 6, 10
direct generation, 32, 109, 129, 152, 173, 189
discretization, *43*, 109, *111*, *115*, 129, *131*, *134*, 152, *153*, *156*, 173, *173*, *177*, 189, 192, *197*
dissimilar, 5, 144
dissipation, 17
distortion, 2
distributed arc heat, 22
double integration, 52, 54
downhand, 10, 13
ductility, 2

E

egen, 32, 109
electrical energy, 5
electric arc, 2, 5, 7, 15, 17
electric discharge, 5
electrode, 5–15, 35
    feed rate, 7–8
    melting rate, 8
    stub, 11
element, 3, 25–28, 30, 32–33, 35, *43*, 44–45,
        46–47, 48, 50–55, *69*, 70–71,
        84–86, 90, *91*, 92, **93–100**,
        101, *102–107*, 109, 126, 128–129,
        130–131, 144–145, *146*, 147, *148–151*,
        152, 167–168, **169**, *170–171*, 173,
        186–187, *188*, 189, *190–191*, 192,
        209, **210–212**, 213, **214**, 215,
        216–217, *219*
    size, 32, 51, 109, 129, 144, *148*, 167, 173, 186,
        189, 209, 213, 215, 217–218
    type, 32–33
elliptical, 35
emissivity, 18
energy, 1–2, 5–7, 17
    density, 35
excel, 45, 56

F

fabrication, 1
field variable, 25–27
filler metal, 6, 8, 13–14
finite element analysis, 25, 32, 50, 52
finite element method, 3, 25, 207
flux, 7, 11–13, 15, 17
    cored arc welding, 15
    shielded welding processes, 7
forced cooling, 18
force matrix, 30
forehand technique, 10, 35, *36*
fusion, 2–3, 5–6, 8
    welding, 5

G

gaseous matter, 5
gas metal arc welding, 15, 55, 71, 90, 126, 144,
        167, 209, 215
gas shielded welding processes, 7
gas tungsten arc welding, 13–14, 85, 186, 209,
        217–218
Gaussian distribution, 36, *37*
grain coarsening, 2
grain refining, 2

H

hardening, 2
heat affected zone, 2, 5
heat capacitance matrix, 30
heat conductance matrix, 30
heat distribution, 35, *48*, *128–129*, 168, *170–172*,
        187, *188–191*, 209, 215, 217–218
heat flux, 17, 29, 36–39, 42, 44–45, 52–54, 129, 152
heat input, 2–3, 10, 13–14, 17, 23, 32–33, 35,
        39–42, 44, 48, **49**, 50–52, 55–56, 69,
        84, **87–89**, **90**, **96–97**, **101**, 109–110,
        126–127, 129, *130–131*, 132, **133–134**,
        145, 147–148, *149–151*, 152, **154**, 155,
        168, **169**, *171–172*, 173, **175–176**, 186,
        187, 188, **194–196**, **210–220**
heat loss, 6, 13, 17–18, 22, 30, 32–33
helium, 13
horizontal, 1

I

impure elements, 2
inclusion, 2, 7
inert gas, 7, 13
in-plane analysis, 39, 49, *50*, 126, 144, 167, 186
integrity, 2, 7
interpolation function, 25, 27
isotherms, 20–21
iteration, 31
iterative, 22, 31

J

joining, 1, 5
joint line, 1–2, 7
joint strength, 1

K

kesize, 32, 109
keypoints, 32, 109

L

lamellar tearing, 2
latent heat, 22
lateral direction, 10, 19, 21
length, 7, 11, 27, 33, 41, 44–45, 49, 54–55, 71,
        126, 129, 152, 167, 173, 187
linear, 25, 27
localized heat, 2, 17
localized plastic deformation, 2
longitudinal, 21, 23, 33
low alloy steel, 2

M

magnetic forces, 6, 22
manual welding, 7–9
material properties, 17, 22, 31–32, 126, **127**, 144,
    167, 187
mechanical properties, 11–15
melting point, 2, 5, 17, 20, 22, 145
metal deposition rate, 7
metallurgical bond, 1–2, 5
metallurgical damage, 1
metallurgical issues, 1
metallurgical phase change, 2
mid-section, 2, 49
molten pool, 5–6, 10, 12, 20, 22–23, 35
motorized carriage, 12
motorized drive, 15
moving heat source, 19
multipass welding, 2

N

natural cooling, 18
Newton Raphson's method, 31
ngen, 32, 109
nodal heat, 44, 48, 51, 55, 71, 85, 90, *130–131*,
    132, **133–134**, 148, 168, *171–172*, 173,
    175–176, 188, **194–196**
nodal temperature, 25–27, 29–31, 110
node, 25–29, 32–33, 39–42, *43*, 44, 48, **49**, 50–52,
    55–56, **57–68**, *69*, 70–71, **72–83**,
    84, *85*, 86, **90**, *91–92*, 101, *102–107*,
    109–110, **127**, 129, *130–131*, 132,
    133–134, **145**, 148, *149–151*, 152–153,
    154, 155, 167–168, **169**, *171–172*, 173,
    175–176, 186, **187**, 188–189, *190–191*,
    192, **194–196**, **210–220**
non consumable electrode, 8, 13
non linear, 22, 31
non uniform heating, 2
numerical integration, 44, 54
numerical technique, 3, 25

O

one dimensional, 25, *26*
operator skill level, 1
overhead, 10

P

penetration, 10, 12, 14
periphery, 36–37
phenomena, 3, 17, 22, 207
phenomenon, 2, 5–6, 22

plastic deformation, 2
polarity, 14
polynomial, 25, 27
postprocessor, 32–33
post weld heat treatment, 1
power of the arc, 7
power source, 5, 8–9, 14–15
precipitation, 2
preheating, 1
preprocessor, 32
pressure welding, 5
process efficiency, 6–7, 13–14, 16–17, 37, 45, 51,
    55, 69, 71, 85, 90, 109, 126, 144, 152,
    167, 186, 209, 215, 217–218
productivity, 1, 12

Q

quadrilateral element, 25, 27, 33, 55

R

radial distance, 36–37
radiation, 6, 12–13, 17–18
rectangular element, 50, 54–55, 71, 84, 101
reliability, 1, 22
residual stress, 2, 19
restraint, 2
reverse polarity, 14
riveting, 1
running characteristics, 12
rutile coated electrode, 12

S

segregation, 2
shape factor, 33
shape function, 25–30, 44–45, 52, 54, 92
shielded metal arc welding, 11–12
shrinkage forces, 2
simplex element, 25, *26*, 44
Simpson's rule, 44, 54–55
slag, 11–12
    metal reaction, 12–13
softening, 2
software, 3, 32
solidification, 2, 5
solid modeling, 32
solid phase welding, 5
solution, 3, 32–33, 207
spatial distribution of heat, 19
spatter, 6, 12–13
specific heat, 17, 22, 29, 31–32, 109, 126, **127**,
    144, 167, 187
spreadsheet, 45, 56

stainless steel, 2, 144, 155
stationary heat source, 19
straight polarity, 14
stray heat flux, 37, 129, 152
stringer bead, 11
submerged arc welding, 12, *13*, 16, 45, 109, 209
subsurface, 35
superheat, 17
surface area, 18, 186
symmetrical, 45, 145
symmetry, 41, *43*, 52, 109, 126, 144, 167, 187

T

temperature, 2–3, 5, 7, 17–23, 25–27, 29–33, 35,
　　　51, 110–111, **127**, 132, **144**, 155, 193
　dependency, 22
　dependent, 31, 126, **127**, 144, 167, 187
　distribution, 21, *22*, 33, *112–114, 116–118,
　　　135–139, 156–161, 177–181, 198–200
　gradient, 17, 21, 23, 32
　profile, 17
thermal conductivity, 17–18, 22, 31–32, 109, 126,
　　　127, **144**, 167, 187
thermal cycles, 3, 17, *20*, 22–23, *53*
thermal gradient, 18, 23, 193
thickness, 2, 6, 17–18, 21, 23, 35, 45, 126, 144,
　　　167, 186, 189, 193
three dimensional, 18–19, 21, 23, 25, 32, 39,
　　　40, 186
time increment, 31, 45, 51, 54, 69–70, 109–111,
　　　126–127, 132, 167, 176, 187, 192
time interval, 20, 21, *22*, 25, 30, 32–33, 38,
　　　48–49, 51, 69, 71, 84, 86, 90, 93,
　　　109–111, 132, 168, 176, 188, 192,
　　　209, 215
time lag, 23, 39, 147
time marching scheme, 31
time step, 25, 31–32, 33, *41–42*, 44–45, 48, **49**,
　　　51, *53*, 54, *55*, **69**, 70–71, **84**, 86, 90,
　　　96–97, 101, 109–110, 126–127, 132,

　　　133–134, 144–145, 147, 153, **154**,
　　　155, 167–168, 173, 186, **187**, 192, 209,
　　　210–212, 213, **214**, 215, **216**, 217–218,
　　　219–220
toughness, 2, 13
transient, 25, 29–33, 207
transverse, 19, 21, *22*, 23, 33
two dimensional, 18–19, 21, 23, 25, 32, 41, 49–50,
　　　126, 167
tungsten electrode, 8, 13, 15

V

vaporization, 6, 17, 22, 35
vertical, 10
voltage, 6–9, 12–17, 35, 37, 45, 51, 55, 69, 71, 85,
　　　90, 109, 126, 144, 152, 167, 186, 209,
　　　215, 217–218

W

weaving, 10–11
weightage factor, 45, 54, 56
weld discontinuities, 1
welding, 1–3, 5–19, 23, 32, 35, 45, 55, 71, 85,
　　　90, *102–107*, 109, 126, 128, 144, 152,
　　　167–168, 186–187, 207, 209, 215,
　　　217–218
　phase, 49–50, 155
　procedure, 1
　process, 1–3, 5–17, 35, 37, *41*, 45, 55, 85, 90,
　　　109, 126, 144, 167, 186, 207, 209, 215,
　　　217–218
　speed, 7, 11–14, 16, 39, 45, 51, 54, 70–71, 85,
　　　90, 109, 126, 144, 167, 186, 209, 213,
　　　215, 217–218
　torch, 13
weld line, 7, 12, 14–15, *20*, 49–51, *53*, *132*, 145,
　　　153, *174*
width, 6, 11, 27, 33, 45, 54–55, 86, 90, 92,
　　　126, 152

Printed in the United States
by Baker & Taylor Publisher Services

Printed in the United States
by Baker & Taylor Publisher Services